D1355629

POCKET ENCYCLOPAEDIA OF
CACTI IN COLOUR
INCLUDING OTHER SUCCULENTS

EDGAR and BRIAN LAMB

Pocket Encyclopaedia of
CACTI
IN COLOUR

INCLUDING OTHER SUCCULENTS

with 326 Photographs reproduced in Full Colour

BLANDFORD PRESS
Poole Dorset

Colour printed by D. H. Greaves Ltd., Scarborough
Text filmset by Keyspools Ltd., Golborne,
Warrington, Lancs.
Printed and bound in Great Britain by
Cox & Wyman Ltd.
London, Fakenham and Reading

CONTENTS

FOREWORD

My friendship with Edgar Lamb started in 1947, when we first exchanged correspondence about our respective collections. Since then, he has visited our Exotic Garden in Monaco, the management of which I held for 34 years, and in 1953 I had the opportunity of visiting the Lambs' Exotic Collection in Worthing, Sussex, when the third Congress of the International Organisation for Succulent Plants Study (I.O.S.) of which I am now President, was held in London.

Upon our very first meeting I realised that Mr Edgar Lamb was both an excellent horticulturist, with specialized skill in cultivating plants, and also a keen collector. He has taught his son Brian how to love these beautiful plants, and he has in turn become an expert and a valuable associate to his father's work.

Horticulture is a science which demands a thorough study of each plant, its cycle, ecological conditions, ways of reproduction etc. so that to grow Succulents which come from hot and desert countries in other climates takes years of study and numerous experiments. My friends, the Lambs, have been very successful in this field, and through the years I have followed the development of their work and collection with great interest. It therefore gave me much pleasure to be invited to write the foreword to this, their latest book in which they deal with all the secrets leading to successful growing. I would give just one word of advice, especially to beginners – do not hesitate to follow each step word by word and you will soon have the pleasure of seeing the beautiful flowers of these most unusual plants.

L. VATRICAN,
1st July, 1969.   Director, 'Jardin Exotique', Monaco.

1. INTRODUCTION

The popularity of cacti is increasing year by year, and with so many people joining the already large ranks of cacti enthusiasts, a pocket encyclopaedia has become a necessity. Although this book is primarily aimed at those new to the hobby, many enthusiasts of some years' standing will find it very useful, not only because well over three hundred species are illustrated, but also because of the chapters on general culture, pests and diseases, methods of grafting, and the special section on growing from seed. This section is accompanied by a number of half-tone illustrations, as a guide to seedling growth, together with two line drawings for a small, simple seed raising bench. Seed raising comes at the end of the book, since we find that most people prefer to start with plants rather than seed, and so need advice on growing these successfully first.

It is very difficult to advise people on how to grow their plants, when there are collectors in almost all the countries of the world, living in a wide range of climatic conditions. Obviously, our main practical experience is gained through growing plants in England, where most species have to be kept under glass for most of the year. These conditions will also apply to many countries in Europe, part of North America and Canada, and many areas in Asia, while in the southern hemisphere they will apply to part of New Zealand, Australia, Tasmania, and the most southerly areas of South America. Even so, there will be a certain amount of variation regarding minimum winter temperatures, therefore the advice we give can be modified slightly in places where greenhouses are not necessary, or are only required for the most tender species. The growing seasons often have to be reversed in the southern hemisphere, and in equatorial regions it can be a problem to know when to rest one's plants, equivalent to a normal winter period in other parts of the world. One chapter will be devoted entirely to growing plants in the tropics, and the use of such things as 'lath houses', methods of protecting plants in the open against tropical hailstorms, and so on.

Nearly all the photographs which appear in this book are of plants in our famous 'Exotic Collection of Cacti and Other Succulents' which has been assembled at Worthing, Sussex, in the south of England, over a period of more than forty years. At the time of writing there are about 8,000 species of cacti and other succulents from many parts of the world in this collection. Of its kind it is outstanding in Europe and visited by a large number of people from many countries every year. We have been able to study the plants which are grown under various conditions, noting their habits, health and flowering, over the years, and the advice given in this book is based on our long experience.

In addition to the 300 colour illustrations taken from our collection, a range of plants filmed in habitat is also shown. Such pictures will be of interest to all readers, but particularly to those living in climates where most—if not all—of their plants can be grown in the open air. They give a better idea of the size which some plants attain, and show what fine, attractive specimens they can eventually make. Some of the photographs in the 'wild' are like a rock garden setting, and no doubt many will be able to arrange similar, pseudo-natural settings in their gardens. These habitat pictures have been taken by various well-known authorities, who have not only lent the slides but have been so helpful in many ways in the past. Our grateful acknowledgements are given on page 211.

For a number of years, we have also been studying the frost resistant species, by means of a lean-to greenhouse which is not heated at all in winter—even though the temperature has fallen to 18°F ($-7.8°C$) within. One chapter is devoted to this subject, as it is of interest not only to those who have to grow varieties under cover all the year round, but also to those in semi-Mediterranean climates, where a certain amount of frost can be experienced in winter, though not to the same degree as here or in other temperate areas. The species mentioned in this chapter can be grown completely in the open in those parts of the world.

Finally, in this introduction, we must briefly define a 'cactus' and a 'succulent'. Cacti, or to be more precise the family of plants

Cactaceae, are perennials usually bearing areoles or spine cushions, and can withstand drought because of their succulent stems or bodies which contain water storage tissues. The word 'succulent' refers to any plant with water storage tissues in its leaves or stems, and nowadays numerous tuberous rooted plants are collected as succulents because the storage tissues in the root—which in some cases can be very large—enable them to survive during drought. However, this makes it very difficult to draw the line between some plants collected as succulents, and other ordinary garden plants such as daffodils, tulips, crocuses, etc., whose bulbs or corms contain water storage tissue. Therefore the word 'succulent' is a very loose term, and includes all cacti, although there are many true succulent plants with thorny stems which to the uninitiated look like cacti. It may be helpful to point out that cacti are New World plants, indigenous *only* to northern, central and southern America and the neighbouring islands. Some cacti have been introduced into parts of the Old World, such as areas of Africa, the Mediterranean and Australia, and have gone wild, which has led many people to believe, erroneously, that these varieties are indigenous to the Old World. There are, however, very many other succulent plants throughout Africa, the neighbouring islands, India, Ceylon and Burma, plus the occasional species in Australia, which are not found growing wild in the Americas. In the Americas too there are some other indigenous succulent plants which have developed along similar lines through a process of evolution. For example, the Agaves of the New World are fairly similar in form to many species of Aloes in the Old World, a true case of parallel evolution.

EDGAR AND BRIAN LAMB.

2. WHAT SOIL MIXTURE TO USE

So much has been written on this subject that if the grower of cacti tried to take all the different theories into account he would hardly know where to start. From forty years' experience, it has been found that the best general soil mixture is based on leafmould, sand and loam. Certain modifications have been made in recent years, so that a mixture of sand and leafmould—plus a few additives—is now used for many of the smaller globular cacti and the slower-growing 'other succulents'.

It is true that only a certain percentage of cacti and other succulents grow in soil containing any leafmould (i.e., humus or decayed vegetable matter), but we have found that out of approximately 8,000 different species within our collection, none has suffered through growing in a mixture containing leafmould. Obviously, as with any group of plants, there will be a few which are more difficult to grow, but we use a small amount of leafmould with these varieties too. Since this book is intended particularly for those new to growing cacti, the more difficult kinds are not included. Our aim is to suggest a simple, easy-to-follow, soil mixture, along with advice about the different types of plants, no matter whether you are growing them in the open air, as in more tropical climates, in a greenhouse or garden frame, or only on a window-sill.

However, bearing in mind that this book will be read in many countries, it is hard to ascertain exactly what constitutes a good average leafmould, a good sand and a good loam. Here are our recommendations.

LEAFMOULD

The best leafmould is that formed from beech or oak leaves, or a mixture of the two. If leafmould is being collected in the woods it should *not* be taken from the surface; dig down a few inches to reach the soil which is fine, showing that the leaves have completely broken down. It may be necessary to sieve out a few tree roots, but the really fine leafmould thus obtained is one of the most important constituents of your cactus and succulent soil mixture.

SAND

The ideal sand is a pit or quarried sand, which has been washed a few times; this is obtainable in some countries, but *not all*. We use a local pit sand, but it is made up of fine, medium and coarse grains, so that it never sets solid, unlike the finer sand used for cementing. If you are unable to obtain such a sand it may be necessary to use a silver sand, or some other fine variety, mixed with grit such as flint or granite chippings, etc. Sea sand is not generally suitable due to its high salt content, but should not be confused with beach sand taken from well above high-water mark. Beach sand may be of a suitable grain size, but should be washed well, or allowed to weather out of doors, where the rain will rid it of any salt.

LOAM

The ideal loam is one which does not set too solidly and harden after rain, although even this type can be used with the addition of extra sand and leafmould. Some people collect soil from molehills, and this is often ideal. In areas where the soil is already of a sandy nature, it may only be necessary to add a small amount of leafmould in order to have a very useful general cactus soil mixture.

A GOOD AVERAGE SOIL MIXTURE FOR THE CACTI AND THE OTHER SUCCULENTS:

By volume   1 part loam
                  1 part sand
                  1 part leafmould

To this can be added the following, easily obtainable ingredients, mainly for the true cacti, in the proportions given:

To a one-gallon mixture	4 heaped teaspoons of bone meal
of loam, sand and	3 heaped teaspoons of gypsum
leafmould—add	1 heaped teaspoon of superphosphate

(N.B. We have used a teaspoon as the measure, since this is usually available. The amount added to the soil mixture does not have to be too exact.)

What will these ingredients do? The bone meal is a slow acting fertiliser, while the other two ingredients are an aid to better flowering, particularly with some of the globular cacti. However, a mixture of loam, sand and leafmould should ensure satisfactory results, and most of the plants illustrated in this book should flower very well without any additives.

In more recent years we have been growing many cacti and other succulents in a soil mixture containing no loam, with excellent results. We have used, as an average mixture, equal parts of leafmould and sand, but with certain species requiring slightly better drainage a higher proportion of sand should be added. So, when referring to the brief cultural data for the species illustrated in this book, we shall suggest the use of the average mixture, whether this is loam, sand and leafmould, *or* just leafmould and sand combined; mention of an extra 25 per cent of sand will only be made where necessary. Bone meal, gypsum and superphosphate can be added to the leafmould and sand mixture in the proportions already given.

The only type of plants requiring extra leafmould are the epiphytic cacti, such as the well-known Christmas Cactus, or Zygocactus, and other similar types, e.g. Schlumbergera, Rhipsalidopsis, Rhipsalis, Epiphyllum, etc.; for these varieties, an extra 25 per cent leafmould can be added. In the wild, epiphytic cacti are often to be found growing in nothing but rotted leaves, in the forks of trees.

We have set out what we consider to be a good average soil mixture, but you may buy plants growing in a soil which does not exactly correspond to this. In fact many firms, particularly in Europe, add a very lightweight substance called *Perlite*, which is better known here as *Loamalite*. The size of the granules enables it to hold moisture rather like a sponge, which makes it a useful addition to a mixture, particularly in hotter climates where moisture is lost from the soil at a much faster rate than in temperate areas. The same substance is also used when growing plants by hydroponics; this is a method whereby the mixture used contains no goodness of its own, but one waters with the required nutrient solution.

Growers living in countries where volcanic gravel can be obtained would be advised to try some of this in their soil mixture, as it can be very beneficial. In fact we have seen many cacti, also Stapeliads, growing in little else but volcanic gravel, with wonderful results. However, it may be necessary to experiment with this because— as with sand, leafmould and loam—even volcanic material can vary considerably.

We have refrained from mentioning pH, which is the method of measuring the acidity or alkalinity of the soil, since we feel that too much emphasis has been placed on the ideal degree of pH at which plants grow, particularly as this book deals mainly with the easy to grow and flowering species. Most of the plants included here will grow in many types of soil, and you will soon find the most suitable mixture for your own particular conditions.

The addition of a layer of coarse sand or gravel on the surface of the soil is advantageous, whether growing in pots, beds or rockeries. Not only does it look nice, but it also prevents the plants from being discoloured by splashes of soil and leafmould when watering; and it is advisable to have a good degree of drainage around the neck of most plants. The sand or gravel layer can be approximately $\frac{1}{2}$–1 inch deep.

CAN ONE GIVE FERTILIZERS OCCASIONALLY?

Some liquid fertilizer can be added occasionally when watering; but it is very important not to overdo it, particularly with newly potted plants. Over-watering can reduce root growth, and can easily make the soil sour, which is apparent when the surface becomes green. The only plants which really benefit from the addition of a liquid fertilizer are the epiphytic cacti, especially when they are massed in bloom. The soil mixture suggested will often be sufficient for 3–5 years, depending on growth rate, but should your plants seem in need of a little liquid fertilizer, your nearest florist can probably provide one of the type used for ordinary house plants.

Some growers may wonder whether peat is a good substitute for leafmould. Although the mixture is similar, it contains little or no goodness of its own, and when a soil mixture containing peat becomes very dry, it is more difficult to get it moist again.

3. WHAT SHOULD THE PLANTS BE GROWN IN?

At present, there is much argument regarding clay versus plastic pots, but these are not the only suitable containers; others will be suggested later in this chapter.

Clay pots still seem to be the most generally used containers, and for the beginner at least they have the advantage of drying out more quickly, so that one is less likely to kill plants off through over-watering. Growers with several years' experience, who have become accustomed to the amount of water required, will soon realise that clay pots do have some disadvantages. Due to their porosity, water is continually evaporating through the sides, and valuable nutrients are consequently lost. This can be clearly shown by growing a number of young plants in one clay pot or pan, when it will be found that those in the centre will be quite stunted compared with those nearest the sides. Excessive water loss can be overcome with clay pots by sinking them partially in sand, gravel or even peat, but the plant roots soon emerge from the drainage hole at the base and root down in the surrounding soil. During the warmer weather, which is usually the growing season, it is ideal never to let the pots completely dry out, as the growth of the plant will be affected. However, when beginning to grow these plants it is perhaps safer to slightly under-water rather than over-water them.

Plastic pots do not lose moisture through their surface, only through the drainage hole, and many people prefer their appearance. They are light in weight and do not collect algae on their very smooth surfaces. Nevertheless, we have found that after a few years' use in the greenhouse, where high summer temperatures occur, some plastic pots become brittle. We have received similar comments on these lines from many of our Exotic Collection members who reside overseas, where greenhouses are not required. There is no doubt that plastic pots do have their uses, particularly for those growing plants indoors, and on window-sills.

We mentioned earlier that there were other containers which could be used. In recent years, we have been trying out a number of methods, including staging beds which are 5–6 in. (approx. 13–15 cms) in depth, lined with plastic. These have proved very successful, either with or without drainage holes, since the plants can spread their root systems, and throughout the growing season the roots never quite dry out. Some plants will grow equally well in pots or beds, but there are many varieties which will only do well if they can spread their roots fully. Another advantage to the staging bed method in the greenhouse is that in winter warm air from the heating used will circulate beneath the beds.

Ordinary wood can be used for constructing the beds, but since wood rots easily, it is a good idea to proof it with one of the many wood preservatives. Certain harder woods such as cedar are highly rot resistant for many years, and it is not absolutely essential to line these with plastic. Staging beds can also be made of asbestos, which will last a lifetime, and roofing felt is an alternative to plastic for the lining.

Normally, we use a soil mixture as mentioned in the previous chapter, but, if desired, a layer of gravel or broken pot can be placed on the bottom as a form of drainage, before filling in with the soil.

For those who wish to make a more permanent bed, built on the floor of the greenhouse, or in the open air as in the tropics, bricks or concrete blocks can be used. This is particularly useful in more tropical climates, where heavy rains can occur, as these beds will drain well and consequently reduce the chance of rotting. Nowadays, it is possible in many countries to obtain cement blocks with a rock-face, which can look most attractive. Natural rock is certainly the best of all materials to use, although—unless it is quarried near you—it is the most expensive method of constructing rockeries, raised beds, etc.

The use of concrete troughs or glazed pottery bowls without any drainage holes is not to be frowned upon, provided care is taken with the watering. The important point to remember when planting up a staging bed, rockery, or even a bowl, is to use plants of similar growth rates. This does not imply that plants from only one genus have to be used, and does not necessarily mean that a true cactus cannot be mixed with some of the other succulents. If you make mistakes in this line, you will soon learn by them, as some species can become very rampant growers, and completely cover up plants of the smaller, more choice varieties.

At the end of this book there is a section devoted to seed-raising, in which you will find one or two illustrations showing young cactus plants of about eighteen months, which have been pricked out into galvanised metal trays barely 2 in. (5 cms) deep. These trays have small drainage holes, and prove to be a very useful method of growing young plants. In fact, some of the smaller globular cacti such as Rebutias, Notocacti, etc., which are ideal for beginners, will grow for many years in trays of this depth and flower exceedingly well.

4. HOW AND WHEN TO WATER

The most frequent question asked by new enthusiasts is, 'How often should I water such and such a species?' Unfortunately, it is quite impossible to give a simple answer, since so many factors must be taken into account. Some of these factors were mentioned in the previous chapter in connection with suitable containers for the plants, but climatic conditions are also important. Those living in tropical climates should water their plants much more frequently, particularly where the atmospheric humidity content is very low. Those growing plants in greenhouses or frames in temperate climates will also use a fair amount of water during hot summer weather, whereas plants kept on window-sills—where very high temperatures are unlikely to be experienced—will obviously require less. Most of the true cacti have a winter resting period, when little or no water is given, although plants growing in centrally heated rooms may need watering occasionally to prevent them from shrivelling or even losing joints, as with some Opuntias for example. With few exceptions, the average cactus plant, if dry, is quite happy in a minimum temperature of about 40°F (4·4°C).

The old idea that cacti need little or no water dies hard, but this is far from true. Cacti may be able to go without water for a considerable period without looking too unhappy, but they enjoy a 'good drink' as much as most other plants. Even so, we find with the many growers we meet each year that there is still a general tendency to under-water rather than over-water. Another mistaken idea is that cacti should only be watered at the base, and never from above; after all, rain falls directly on and around species growing in the wild! Unless plants are very near the greenhouse glass or in the open, it is unusual for them to be burned by globules of water acting as a magnifying lens; the possibility of this can easily be avoided by watering in the early morning or late afternoon. In warm weather the latter is preferable, as it gives the plants ample opportunity to absorb as much as possible before normal evaporation takes place from the surface of the soil or—if a clay pot is used—from the container itself.

Having established that overhead watering is safe, one can use either a watering can or a hose-pipe, equipped with a suitably fine spray. The pressure of water will obviously be much lower when watering pot plants than where plants are grown in the ground or in raised beds, whether under glass or out in the open. Our water in Worthing in the south of England is permanently hard, and does not stain the plants with white chalky marks, which is most likely to happen where temporary hardness occurs. Rainwater is obviously best, but most growers are unable to have enough available without a very large storage system, particularly once their collections enlarge above a dozen or so plants. Growers living in an area where plants are likely to be marked by this chalky deposit can fit′ a filtration plant, such as an ordinary household water softener Advice on this is best obtained from someone with local knowledge, such as your nearest Parks and Gardens Superintendant, who will usually be pleased to help.

It may seem that we have still not given any real indication as to how much water and how often! Briefly, then, remember that in warm growing weather the root-ball should not quite dry out; if you give the plants too much water this will soon show, as the surface of the soil will probably go green in patches through becoming sour. Here is a rough guide to watering as we do it, bearing in mind that when pot plants are watered overhead, sufficient should be given to fill the pot to overflowing.

WATERING GUIDE IN THE EXOTIC COLLECTION:

Plants growing in clay pots Once every	*Spring and Autumn* 7–10 days	*Summer* 4–5 days

This is assuming a maximum greenhouse temperature of 70–80°F (21·1–26·7°C) in the spring and autumn, and over 90–95°F (32·2–35°C) in summer. If plastic pots are used, plants can perhaps be left a day or two longer before watering again. It is also quite good practice, if time allows, to give them all a light spray over daily or every other day during warm weather. Plants growing in raised staging beds or in the ground can often be left twice as long between waterings, since the soil will lose its moisture at a much slower rate. When we water these sections of our collection, we always do so very thoroughly, so that the moisture will penetrate down to a good depth and the more deeply rooted species will benefit as much as those with surface roots.

Our advice on watering has assumed that all plants require water from spring to autumn, but as will be seen in the section following the colour illustrations, there are a few groups with a slightly different growing period such as Lithops and Conophytums. A special cultivation note will be found for these species, and also for the epiphytic cacti such as Zygocactus, or Christmas Cactus, which require some water throughout the year. In case we have given the impression that all cacti, including the other succulents, need the same amount of water, we should point out that this is not quite true; however, in order to facilitate easy watering extra sand can be added to the soil, which will improve the drainage. A note of this will also be found in the section following the colour illustrations, where applicable.

For those growing plants indoors, the watering table can still be used as a guide, but if the pots are standing in small dishes, it is quite easy to see when the last trace of moisture disappears from underneath the pot. Where there is central heating in winter, it is advisable to give some water at intervals of perhaps 3–4 weeks or less to avoid shrivelling, even if this means that the plants continue to grow quickly. Experience should soon enable you to judge the amount required, and to avoid those species needing a winter resting period. Plants growing in troughs or table gardens, whether there is drainage or not, can be watered with roughly the same frequency as those in staging beds.

The various modifications for watering necessary in tropical countries are given in Chapter 7.

5. PESTS AND DISEASES

As with all types of plant life, there are always pests and diseases to contend with, but, fortunately, if your cacti and succulents are given reasonable care, there should not be too much trouble. The chief pests are:

Mealy bugs   which can affect the aerial part of the plant as well as the root system.

Scale insects   like small limpets on the body of the plant.

Red spider   minute red insects which attack the body of the plant.

There is also, to a lesser extent, root eel worm, which is rare in the United Kingdom, and occasionally the common-or-garden green-fly and blackfly.

It is difficult to suggest possible insecticides to be used, as the trade names often vary in different parts of the world. We will mention a few, also some simple methods of killing these pests, which are particularly applicable to those with small collections; but it is a good idea to contact a well-known firm of insecticide manufacturers in your own country. One can often get the desired information by visiting a local florist specialising in general house plants, and instructions about suitable insecticides are usually supplied with these.

For those with a very small collection, it is possible to look over the plants one by one, and if any mealy bug or scale is detected it can easily be removed using a small artist's brush with a solution of household soap or methylated or surgical spirit. If the latter is used, the plant should be put in the shade for a day and sprayed with water before being returned to its normal position in the greenhouse, conservatory or window-sill. However, it becomes a little more difficult to use this method once you have a small greenhouse full of plants, and an insecticide which can be watered or sprayed on is then required. Such insecticides can be divided into two groups: contact poisons, which as the term suggests kill the insects on contact with them, or systemic insecticides. The latter, which are more deadly and need to be handled with great care, are absorbed into the plant; pests feeding on the plant juices also ingest some poison which kills them.

A good contact insecticide is pyrethrum available in suspension as liquid which can be easily mixed with water, and applied as a drench. When used in this way it will not only kill mealy bugs and certain other pests on the plant, but also in the soil as with root mealy bug. Mealy bug can certainly be kept down if this is used about six times a year during the spring to autumn period. Pyrethrum, as with many other insecticides*, can also be obtained in aerosol can form. This method of usage can be very useful indeed if pests appear during the dry resting period, when it would be harmful to apply an insecticide in drench form.

Insecticides in aerosol form are not recommended for use in a greenhouse for fear of breathing the dangerous fumes. We cannot over emphasise the need for instructions on cans to be read very carefully. If ever you have problems with obtaining a suitable insecticide, you cannot go far wrong by purchasing one that is recommended for use on general 'house plants'. If ever in doubt about the dilution required it is reasonable to use the insecticide at a strength recommended for house plants. The plants most likely to be affected are Kalanchoes, Crassulas, etc., but in our experience the true cacti do not seem to be harmed provided spraying is done in the cool of the evening; in fact this is the best time to do any spraying, no matter what type of insecticide is being used. With a big collection, it is a good idea to spray with a different insecticide throughout the growing season at intervals of 6–8 weeks, and develop a rotation system using three or four different kinds. Despite the poisonous nature of these products, pests can develop a resistance to them, but this is unlikely if you 'ring the changes' rather than persist with only one make.

Red spider can spoil plants in a very short time, and disfigure them badly almost as though they have been burned. Chamaecereus, Coryphanthas, Rebutias and other rather soft-skinned cacti are

---

* Malathion, metasystox, dimethoate.

particularly prone to this insect, which enjoys dry conditions. However, if the plants are given ideal growing conditions—including a certain amount of humidity in all warm weather—this pest is rarely found. The periodical use of nicotine (using the strength advised on the can) is very effective. Following an infestation, three sprayings may be necessary at fortnightly intervals, in order to kill off all the insects as they hatch from their eggs.

Root eel worm, or Nematode as it is sometimes called, is rare in the United Kingdom, but more common in southern Europe and parts of the United States, particularly on the western coast. This small pest will distort root systems badly, and so make the plant very unhealthy. It is more difficult to kill, and surrounding soil, pots, etc., must be sterilised. Systox will kill it, but should it be suspected, a local firm which manufactures insecticides should be consulted for advice.

The ordinary greenfly and blackfly occasionally turn up in the greenhouse, particularly where flowers, flower spikes or even certain plants exude a sticky juice or nectar. Spraying with soap solution is often sufficient to kill these pests, without resorting to one of the stronger insecticides.

There are not many true diseases which attack cacti and succulents. Damping-off disease only affects young seedlings as a general rule, and is dealt with in the special seed-raising section at the back of the book. Orange rot occurs in a few American cacti, particularly certain Ferocacti, where it usually starts at the root; the only solution is to cut it out, dry the plant off and re-root, treating as for 'cuttings' (see Chapter 9). This kind of rot is not to be confused with ordinary rotting caused by over-watering, but the treatment is the same. Orange rot can also follow on a bad attack of root mealy bug, which can occasionally start the rot from the root upwards.

There is also a black-rot which mainly occurs in Stapeliads, and the treatment is similar to that for damping-off disease, except that a double strength solution can be used as an overhead spray or for painting on the affected parts. Black-rot is sometimes difficult to kill, but the best preventative is to use the potassium hydroxyquinoline sulphate solution as a regular spray about three times

north–south direction, which should allow the sun to reach every corner, but this is not of course absolutely essential.

Ventilation is also very important, but care should be taken not to overdo this, otherwise the greenhouse becomes a 'wind tunnel'! If this occurs, the plants will probably be losing moisture faster than their roots can absorb it from the soil; such rapid transpiration is of little benefit. Even in hot weather we rarely open any ventilators, and find that the opening of doors at one end is usually sufficient except on the rare occasions when there is a real heat-wave. Our greenhouses vary from 50–150 ft (about 15·2 to 45·7 m) in length, therefore the opening of a door at one end should adequately ventilate a small greenhouse. If ventilators are preferred, it is better to fit one on each side of the greenhouse, rather than on the roof, since roof ventilators often leak, and side ventilation is just as effective.

Although some people say it is not necessary for greenhouses to be artificially shaded in climates such as that in England, this is far from true. Not all plants need shading, but the majority of those illustrated in this book will grow much better if the glass is lightly shaded. The easiest and cheapest method is to spray or lightly dab some whitewash on the outside of the glass. This is best applied with a garden syringe, so that the sun is not sompletely excluded, and if some cheap-grade unused car oil is added, then the mixture will not wash off too quickly. We suggest the following proportions in an average 2 gallon (4·5 litre) bucket:

> 6–10 handfuls of powdered whitening (or more according to density required).
> 1–2 cupfuls of oil.

This should be mixed with a little water into a thick paste, and then diluted until it can be sprayed on easily. It will be necessary to give a repeat application during the summer months, unless no rain at all is experienced, but in the autumn it can be allowed to wash off naturally.

Once the greenhouse has been built and used for its first summer season, it is necessary to decide how to heat it in winter, assuming that you are in a country where frosts occur during this period. There are four methods of heating:

(1) a boiler system where the water pipes are heated by coal, oil or gas;
(2) electric fan heater or tubular heater;
(3) flued gas convection heater;
(4) small oil heater.

The average collector will probably consider either some form of electrical heating, or a small oil heater which needs refilling with paraffin every few days. An oil heater will be cheapest, and provided the wick is kept clean it is very reliable; although some extra humidity will occur through the burning of paraffin, it is not likely to harm the plants. If electricity is preferred, we would suggest a fan heater, as this provides a better heat distribution around the greenhouse, and there is little chance of stagnant pockets of air in the corners, as can happen with tubular heaters. A thermostat will reduce costs, and for the majority of plants this can be set for a minimum of 40°F (4·4°C). Since electricity can fail at times, and is most likely to do so on a really cold night, it is advisable to have a paraffin heater as a safeguard. In fact the ideal system for the amateur is to use a paraffin heater most of the time, with an electric fan heater in reserve to boost it in severe weather. Not only is this an economical and reliable method, but growers who have to go away or fall ill can safely leave the electric fan heater to do the job by itself.

We do use oil heaters for one or two small greenhouses, but have found large flued gas convection heaters very satisfactory in the main. The chief advantage is that within 10–15 minutes of starting from cold the greenhouse is up to the required temperature. Another advantage is that since greenhouse air is taken in to aid combustion but only heated, dry air is given out, humidity quickly disappears. Until recently such heaters were only manufactured in larger sizes, but we believe one or two smaller models are available now.

The same heating methods can be used where collectors are

restricted to a garden frame, but obviously a much smaller heater will be sufficient. For example, a paraffin heater of the type used under cars in winter, would be quite adequate for a garden frame.

We mentioned in the Introduction that it is possible to grow many plants in a cold greenhouse, and these are dealt with in Chapter 10.

As the question of suitable containers for plants has already been covered in some detail, it may be opportune to mention several types of greenhouse stagings. A wooden, slatted staging is good in that air circulation in winter allows warm air to filter up between the pots. However, growers with more than a few dozen plants will find that a solid staging is better, since this provides more space. Some collectors may wish to keep their plants in pots, yet retain the advantages of a staging bed; this can be partially achieved by bedding the pots into gravel, sand, etc., in which case low sides to the staging will be required. The surface of the staging can be anything from asbestos sheeting or metal, to plastic or even roofing felt. Allowance for drainage of excess water is a matter for each individual, but may be advisable for new collectors.

7. GROWING IN MORE TROPICAL CLIMATES

Some of the information already given in this book will be of use to collectors in all parts of the world, but certain modifications will obviously be necessary with variations in climatic conditions. In tropical countries a greenhouse—as we know it—will only be required for either the most tender species or for seed-raising, but in some such areas hailstorms may occur, and it is important to have a layer of fairly fine, but strong, chicken wire over the exterior of the building. In temperate climates, hailstones which are large enough to break glass are rare, but they are not so uncommon in other parts of the world. Growers in different countries must be guided by local conditions.

In tropical areas which do not usually have long periods of rain, it may be necessary to erect lath-houses to give some shade to the plants. As the name implies, these are simple basic structures on which rows of wooden slats are fixed; gaps between the slats ensure that as the sun passes across during the day, it casts a moving shadow. One lath-house is usually sufficient for many plants, and some species which do not like too much sun will have to be grown in this way permanently. Young plants started from seed will also need this treatment, as will new specimens which may have been grown under shaded glass prior to purchase for instance. Even strong Ferocacti burn if transferred from shaded glass to the open air, but a few months in a lath-house will be sufficient to acclimatise them to your own conditions.

Other tropical countries with very high winter temperatures may have a month or so when long periods of rain occur. Where this is likely, it is best to grow the plants in raised beds, in much the same way as suggested for the greenhouse, but giving them improved drainage by using extra sand and gravel. A simple framework covered with a layer of thick-grade polythene plastic can be erected over such beds, or this can be made much stronger by placing the polythene sheeting between chicken wire, thus it will stand up to heavy rains and even average hailstorms. Clear polythene is quite inexpensive, but may only last one season, as ultra-violet light tends to perish it rather quickly in the tropics. Any structures erected over raised beds should be open-sided to allow free air flow, but corrugated vinyl sheeting or even corrugated fibreglass, can be used if preferred. The latter, though more expensive, is very strong, and will still allow sufficient light to penetrate. It may be interesting to note that some growers use thick-grade polythene to double-glaze their greenhouses in areas where severe winters are experienced, or even to avoid the need for any means of heating where frosts are only slight.

In Mediterranean climates, or other places such as the southern States of North America, many plants can be grown in the open, but where there are occasional slight frosts a greenhouse or open-sided glass lean-to rather like an open verandah will be required. This is usually sufficient to prevent frost damage, provided the plants are dry.

The seasons are completely reversed in countries south of the equator, so when we refer to the growing season for Lithops and Conophytums, this must be altered by approximately six months.

There are countries with almost constant high humidity, which enables plants to grow throughout the year, but it is then very difficult to rest them as we do in winter in temperate climates. Some of the slowest growing plants will absorb sufficient moisture from the soil without actually being watered, though certain of the stemless Mesembryamemums tend to split under such conditions, which spoils their general appearance.

It may be possible for some growers to use trees to provide shade for plants requiring it, but this is not a good method in periods of excess rain, when water can come through in torrents and either damage the plants or wash the soil away.

Finally, insect pests such as ants must be expected. These can cause considerable damage, but most of the more lethal insecticides now available should enable growers to deal with the problem effectively.

It is not possible in a book of this size to cover every point which may arise, but this chapter should give a guide on certain modifications to our own method of growing in the United Kingdom.

8. THE REFERENCE CALENDAR

This chapter is mainly for those growing plants in the northern hemisphere under glass, where climatic conditions are similar to those in Britain. There will obviously be a considerable variation in the degree of cold experienced in winter, which will mean minor adjustments to plant treatment. Where winters are very severe, as in many parts of Scandinavia, Canada and some states of North America, this will result in a shorter growing season, altering the care of plants by a few weeks at least. For those in the southern hemisphere who also need to use greenhouses, everything should be reversed by approximately six months.

JANUARY

There is little greenhouse activity at this time of year, other than the flowering of some of the epiphytic cacti such as *Zygocactus truncatus, Schlumbergera bridgesii* and a few others. The true *Zygocactus truncatus* will have flowered in November and December, only the later forms flowering now. As the plants bear many flowers they may need watering at least once a week, but this will depend on the temperature of the greenhouse. They enjoy a winter temperature with a minimum of 50°F (10°C), but still flower in lower temperatures when less water will be needed and the flowering period may be a little later. Flowering is also likely to be delayed if Zygocacti are grown in an excessively high winter temperature. Certain succulents can be given a little water occasionally, but this should be done during milder weather, assuming a minimum greenhouse temperature of only about 40°F (4·4°C) is normally maintained. Such watering should be restricted to the leaf. succulents; plants such as Conophytums and Lithops should be kept quite dry.

FEBRUARY

For the first part of the month treatment should be as for January, but more sunshine will often occur later, when one or two of the globular cacti—including Rebutias and even a few of the Mammillarias—may show signs of flower buds. These plants, particularly the Rebutias, will enjoy being watered if weather permits, so that they start to fill out after their winter rest. Many varieties can be re-potted now, provided they are only being transferred—with extra soil—to a pot with a wider diameter, and the root ball is not disturbed. If, however, you find a plant which has root mealy bug, put it back into its pot immediately and separate it from the rest. It can be unpotted, and the roots washed with an insecticide, in late March or April, then allowed to dry in the air for a few days before re-potting. All plants should be checked for mealy bug, and any found on the plants themselves can be dealt with by applying one of the insecticides, or even methylated spirit, with a small brush. If spirit is used, remember to spray with clean water a few hours later, before allowing the sun to shine on affected specimens.

MARCH

Many plants will now be showing growth, and most of the leafy succulents will be needing more water; if the weather is good, with plenty of sun to warm the greenhouse, many of the globular cacti and some of the larger epiphytic cacti will also be in bud. In this case, the appearance of your plant will be greatly improved by a weekly spray, to wash off the dust which has collected during the winter. The first flowers on the globular cacti can be expected this month, and Rebutias should have rings of flower buds showing in addition to the occasional fully developed bud or flower. The larger epiphytic cacti, which produce sizeable flowers, will require more water and should also be given an occasional feed with any fertilizer recommended for house plants, such as Bio.

APRIL

All plants except some of the stemless Mesembryanthemums (such as Conophytums and Lithops) can now be given a good watering; this is best carried out early in the morning on a day expected to be warm and sunny, since nights may still be chilly. A little ventilation on one side or at one end of the greenhouse may be necessary, although the plants will enjoy a humid atmosphere. When the first thorough watering has been completed, it is advisable to raise the temperature for a night or so to around the 50°F (10°C) mark if possible. Within a few days the Opuntias, many of which tend to shrivel during the winter rest period, will fill out and show life. Many spring flowering plants will be in bud now, and, given a few days of sun, will make rapid progress. The stemless Mesembryanthemums can be sprayed lightly from now until June, sufficient to keep the roots alive, but not enough to make them grow. It is essential that the old leaves or bodies should shrivel completely before normal watering is commenced.

MAY AND JUNE

An ever-increasing amount of growth and flower will be observed, and with higher temperatures increasing amounts of water will be needed. As mentioned earlier it is impossible to state precise quantities, but new growers will soon gain experience, and as long as the soil in the pots or beds is not becoming sour, progress can be regarded as satisfactory.

JULY, AUGUST AND SEPTEMBER

Growth on all plants will continue. You can now begin to water the stemless Mesembryanthemums (chiefly Conophytums and Lithops) in the same way as the other plants, and buds can be expected on any of them within a few weeks. If the occasional plant has not dried off completely, it is perfectly safe at this time of year to peel off these leaves, but do not water overhead for a day or so after this has been done. It is advisable to remove the dried dead leaves or skins from all plants in this group, otherwise mealy bugs can find very convenient homes there. Leafy succulents, such as Echeverias and Kalanchoes, should have developed fine heads of new leaves, but if their water supply is now halved, they can be expected to develop the highly coloured leaves which make them so attractive. During this hottest part of the year red spiders are much in evidence, and can spoil plants very rapidly, so it is important to watch out for these pests; plenty of overhead spraying is a deterrent. Shading of some sort will be needed from July onwards, although in many areas such as Worthing, Sussex, it is sometimes required as early as May. Towards the end of September, those who experience more severe winters should start reducing their watering, whereas it is not usually necessary for the rest of us to do so until October. However, the early onset of autumn, even in milder areas, can sometimes mean that watering should be reduced in frequency in September, except for the stemless Mesembryanthemums and the leafy succulents.

OCTOBER

No more shading will be necessary, and this can usually be left to wash off naturally with the rains. Watering of most globular cacti can be stopped, except for very late flowering species, but a good show of flowers can still be expected on the stemless Mesembryanthemums, Stapeliads, etc., and the first flower buds should be showing on *Zygocactus truncatus*. Dead leaves should be removed regularly, as they can cause mildew and rotting at this time of year. Little or no ventilation will be needed now. Greenhouse heating equipment should be checked over, so that it is ready for use at any time.

NOVEMBER AND DECEMBER

Except for a few succulents, such as some of the Crassulas, Anacampseros and the stemless Mesembryanthemums, little flower will be seen from now on except for the epiphytic cacti, which as they come into bloom can be transferred to the house for decoration purposes. At the end of one year you will have learned a great deal, purely through your own practical experiences.

9. VEGETATIVE PROPAGATION
and
SIMPLE GRAFTING TECHNIQUE

For various reasons it is a good idea to have one or two spare plants of each species, either to replace an original specimen which has died, or to give to a friend who is perhaps developing an interest in cacti and succulents. Obviously, if you have grown plants from seed, you will be more likely to have a few spares of each species, but where plants have been bought this will not be so.

Many true cacti, as well as the other succulents, branch freely, so that it is easy to take some cuttings, preferably cutting these off at the narrowest point. With true cacti it is better to leave such cuttings for about 10–14 days, until they have calloused sufficiently, before rooting them up. However, if the cut surface is very small, only a few days need elapse before planting. On the other hand, if the cut has to be made at a broad position, as with many of the larger Cerei and Euphorbias, it may be safer to leave the cuttings to callouse off for three to four weeks, during which time they should be left in a warm dry position, but shaded from direct sun. Such treatment is also necessary if a plant rots at the base during the winter, and has to be cut off just clear of the rot. Many of the leafy succulents, also the 'leaves' of epiphytic cacti, need only be left for two or three days before planting; cuttings which are left too long may shrivel and be too weak to take root.

In addition to taking cuttings by removing offshoots, there are other methods of vegetative propagation. Some succulents, especially many of the leafy kinds, can be propagated by carefully removing a perfect leaf, placing on sand or a suitable soil mixture, and lightly spraying every day or so. With luck these leaves will first take root and then produce a new shoot, which will soon grow into a perfect plant. The same method can be adopted with many species of Echeverias, Graptopetalum, Sedums, Kalanchoes, Bryophyllums, etc., although the last-named genus forms young plants automatically on its leaves, which drop off and grow independently. Some, if not all, Gasterias, Haworthias and other plants can be propagated by cutting the leaves up into a number of sections and treating each as an individual leaf. However, as these leaves are more succulent, they should be allowed to callous for a day or so before planting and spraying. The well-known Sansevierias can be treated in this way very successfully.

We have mentioned sand, etc., as being a suitable substance in which to place cuttings, and we invariably use a mixture of sand and leafmould, with a higher proportion of the former. This means that the cuttings can be left in the tray or pots in which they have been placed, since they are in a medium which provides nourishment. For some of the more difficult or slower-rooting varieties, vermiculite or loamalite (Perlite) is very useful, as it contains a fairly constant amount of water; this also applies to many Euphorbias and certain of the rare slow-growing cacti. So far as the use of rooting hormones is concerned, we have carried out many tests and have found that they make little difference to most of the true cacti and other succulents. However, they are of assistance for some of the very slow-rooting species, and we particularly recommend medium strength Seradix in powdered form. Growers can often carry out very useful experiments of their own when trying these preparations. Remember that all cuttings need a warm, shady position in order to root.

Many cacti and some of the other succulents rarely branch, and the usual means of propagating them is from seed. It is, however, possible to decapitate a plant, about half way up the stem, dry off the top part and re-root; with luck the old base will then produce a number of branches, which can later be removed and rooted. Before taking such drastic action with one of your most prized items, however, it is wise to try an experiment on one of the more common varieties. At this point it may be helpful to introduce you to a simple method of grafting, which consists of placing the cut surface of one plant on that of another, so that the two cut sections can unite and grow as one plant. It is usual to put a slow-growing species onto a strong, quicker-growing stock, in order to raise a large plant more speedily. The same process can be adopted to obtain a number of cuttings, since some plants do not normally branch freely on their own roots but will do so when grafted. In addition to using grafting as a means of getting surplus cuttings, or growing a plant very quickly, it can only be carried out in order to save a valuable plant which has lost its roots, or to grow certain cristate or other species which are difficult to cultivate on their own root systems. For example, many cristate plants form into tight mounds, and when grown on their own roots in a damp humid winter atmosphere, rotting can occur because there is insufficient air circulation around the plant. This can be avoided by growing such a plant on a graft. A few illustrations of grafted plants are shown in colour, Plates 55, 124, 135, 187.

A plant can be grafted at any age, although it is easier to carry it out on a young specimen which has not yet become woody. It is a very simple matter to graft young plants between one and two years of age, but it can also be done with very young seedlings. All one needs is a very sharp knife or razor blade, plus a steady hand. A page of diagrams follows this chapter, with an accompanying key describing the various stages. This is more or less self-explanatory, and it is only necessary to mention two other points: (a) the stock to use, and (b) the after-care of freshly grafted plants.

Any cactus can be grafted onto another, but a strong growing stock has obvious advantages. We can highly recommend the following: Opuntia pads (thick varieties with as few spines as possible), also various Trichocerei, Cerei, Myrtillocactus geometrizans, Pereskia, Pereskiopsis and Hylocerei. When grafting the other succulents it is essential to use plants of the same family, but it is probably true to say that at this stage most of you will be more interested in grafting cacti than other succulents.

Once plants have been grafted, the stock should be kept moist by watering at the base, and the scion—the new top of the plant—should be kept shaded and in a warm position for 10–14 days. At the end of this period the fixings or weights can be removed, and provided the scion does *not* drop off the whole plant can then be treated normally. However, if the stock used is not an Opuntia,

Trichocereus or Cereus, it will be necessary to winter the plants in a minimum temperature which does not drop too far below 50°F (10°C). There are a few Opuntias which are tender, but by far the majority likely to be grown will stand fairly cool winter conditions.

DIAGRAMS ILLUSTRATING THE GRAFTING TECHNIQUE

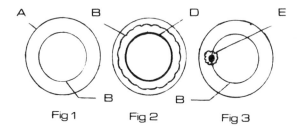

Fig 1   Fig 2   Fig 3

Fig. 1. Top view of stem, which has already been cut across to reveal the vascular ring.

Fig. 2. The same cross-section, but much of it has been obscured by the scion which has been placed centrally upon it, as the scion and the stock were near the same diameter.

Fig. 3. Shows the method used when a small scion is placed on a large stock, thus allowing the vascular tissues of both scion and stock to join.

A = Epidermis of stock.   B = Vascular tissues of stock.   C = Epidermis of scion.   D = Vascular tissues of scion now superimposed upon B.
E = Small scion on part of B.

Fig. 4. A side view prior to scion being placed upon stock 'centrally':

Fig. 5. The same a moment later when they have been placed together.

F = Soft tissue exposed through epidermis having been cut away on the angle.
D = Vascular tissue of scion.
B = Vascular tissue of stock.
G = Soft tissue of stock exposed through epidermis having been cut away on the angle, in the same way as was done for the scion.

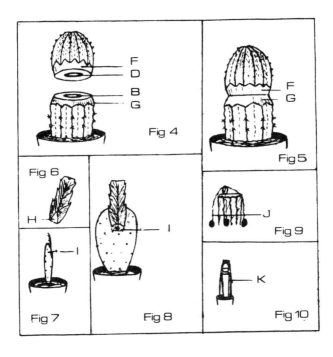

Fig. 6. Piece of an epiphytic cactus such as a Zygocactus, with a small piece of the epidermis at the base cut away to expose the vascular tissue.

Fig. 7. Zygocactus leaf secured by a spine, having been inserted into a slit made. in the Opuntia pad just above an areole. This is a side view, Fig. 8 shows the same, but looking at the flat surfaces of both Zygocactus leaf and pad.

H = Cut surface of Zygocactus leaf.   I = Spine for securing Zygocactus leaf to Opuntia pad.

Figs. 9 and 10. Two methods of holding scion to stock. Fig. 9 is using string and weights suitable for grafting on a large stock. Fig. 10 demonstrates the use of a wire hoop, pressed down into the soil on each side of the stock.

J = String with weight.   K = Wire hoop pressed into soil.

10. WHAT CAN BE GROWN IN AN UNHEATED GREENHOUSE?

About five years ago, when improving and extending the Exotic Collection, we decided to use one of our older greenhouses, which is a lean-to approximately 120 × 12 ft (36·5 × 3·6 m), as an unheated greenhouse. It runs north to south on the western side of an old, high brick wall, and is now arranged in a scenic layout with winding paths. It might not be thought that in Worthing, in the south of England, the weather is ever sufficiently severe for our experiments in this way to be of value. However, we have recorded temperatures as low as 0°F (−17·8°C) in the outside garden, and during the winter of 1962 we experienced three months of freezing weather, in which time the temperature only rose above freezing for an hour or so on two days! Throughout this particular winter the night temperature in the unheated greenhouse often fell to 20°F (−6·7°C). It is true that we do not get much snow in Worthing, but snow can be a blessing for the unheated building, since it insulates to some extent against lower temperatures penetrating from outside.

Apart from the suitability of our conditions, there are positive advantages to such an experimental project. Whereas the amateur can heat one greenhouse without undue difficulty, heating two greenhouses is quite a different proposition. This need not be a restriction, however; in addition to some cacti and other succulents which can be cultivated out of doors in winter, there are many more which can be grown in a dry, unheated greenhouse. Opuntias, which many amateurs like to see in flower, unfortunately tend to grow large and occupy too much valuable space in a small greenhouse. Yet many of them are frost-hardy, provided they are reasonably dry—particularly around the neck of the plant—and will show appreciation of a free root-run by flowering in great profusion. There are many growers who will not mind spending money on a greenhouse or lean-to, if they know that there will not

SOME FROST-RESISTANT CACTI

Cereus peruvianus
,, variabilis
Echinocereus chloranthus
,, coccineus
,, fendleri
,, rosei
,, stramineus
,, triglochidiatus
,, VIRIDIFLORUS
Echinopsis eyriesii
,, multiplex
Neobesseya (any species: MIS-
SOURIENSIS is the most hardy)
Opuntia alcahes
,, bergeriana
,, CANTABRIGIENSIS
,, chlorotica
,, COMPRESSA
,, crassa
,, DRUMMONDII
,, durangensis
,, ENGELMANNII
,, erinacea
,, ,, var. ursina
,, GRANDIFLORA
,, herrfeldtii
,, hickenii

,, JUNIPERINA
,, kleiniae
,, leptocaulis
,, lindheimeri
,, linguiformis
,, mackensenii
,, macrarthra
,, ovata
,, PHAEACANTHA (and vars.)
,, RAFINESQUEI
,, ramosissima
,, rhodantha (and vars.)
,, robusta
,, rufida
,, russellii
,, santa-rita
,, scheeri
,, sphaerica
,, spinosior
,, stanlyi
,, whipplei
Trichocereus bridgesii
,, candicans
,, ,, var. gladiatus
,, macrogonus
,, spachianus

In addition to these, we have found that many other species are capable of withstanding much cold, including the occasional Lobivia, Rebutia, Mammillaria, etc.

SOME FROST-RESISTANT SUCCULENTS

Agave americana (and all vars.)
,,   bracteosa
,,   filifera
,,   lophantha
,,   parrasana
,,   PARRYI
,,   parviflora
,,   stricta
,,   toumeyana
,,   UTAHENSIS
,,   ,,   var.
                    NEVADENSIS
,,   victoria-reginae
Aloe arborescens
,,   aristata
,,   brevifolia

,,   distans
,,   mitriformis
,,   X spinosissima
,,   zebrina
Crassula anomala
,,   lycopodioides
,,   SARCOCAULIS
Euphorbia cereiformis
,,   heptagona
,,   marlothiana
,,   resinifera
LEWISIA (various species)
Ruschia karooica (and at least two other species)
Sedums (numerous species including many which can be grown out of doors)

CONCLUSION

Growers will also be able to experiment with surplus spare plants of many other species, thereby adding more plants to their unheated greenhouses. This may be brought about by using spare seedlings, or by rooting up extra cuttings for the purpose. It is important to remember, however, that with some species which have a very wide distribution, it is sometimes only the forms from the more northerly or alpine locations which are likely to be really frost-hardy.

11. SIMPLIFIED CLASSIFICATION OF GENERA

following the system used by the Authors in their series *The Illustrated Reference on Cacti and Other Succulents* (Blandford).

Family. CACTACEAE

Group	Genera
Opuntia	Opuntia
	Pereskia
	Pterocactus
Cereus	Borzicactus
	Browningia
	Carnegiea
	Cereus
	Chamaecereus
	Cleistocactus
	Echinocereus
	Harrisia
	Lemaireocereus
	Monvillea
	Myrtllocactus
	Neoraimondia
	Pachycereus
	Selenicereus
	Stetsonia
	Trichocereus
	Weberbauerocereus
	Wilcoxia

Group	Genera
Pilocereus	Cephalocereus Espostoa Haageocereus Oreocereus
Echinopsis	Acanthocalycium Echinopsis Lobivia Rebutia Soehrensia
Echinocactus	Ariocarpus Astrophytum Copiapoa Echinocactus Echinomastus Ferocactus Frailea Gymnocalycium Homalocephala Melocactus Neoporteria Notocactus Parodia Stenocactus (Echinofossulocactus) Thelocactus Toumeya
Mammillaria	Coryphantha Dolichothele Epithelantha Mammillaria Neolloydia Pelecyphora

Hatiora & Rhipsalis Epiphyllum & Phyllocactus	Hatiora Disocactus Epiphyllum Nopalxochia Schlumbergera Zygocactus	
Family	*Group*	*Genera*
Agavaceae (formerly Amaryllidaceae)	Agave	Agave
	Sansevieria	Sansevieria
Apocynaceae Asclepiadaceae	Pachypodium Stapelieae	Pachypodium Caralluma Duvalia Hoodia Huernia Stapelia Stultitia
	Ceropegia Hoya	Ceropegia Hoya
Crassulaceae	Sempervivum	Aeonium Greenovia Monanthes
	Cotyledon	Adromischus Cotyledon
	Crassula Echeveria	Crassula Echeveria Graptopetalum Pachyveria
	Kalanchoe Orostachys Sedum	Kalanchoe Orostachys Sedum
Compositae	Senecio (Kleinia)	Senecio

Family	Group	Genera
Euphorbiaceae	Euphorbia	Euphorbia
Ficoidaceae	Mesembry- anthemum	Astridia Conophytum Glottiphyllum Hereroa Imitaria Lampranthus Lapidaria Lithops Ophthal- mophyllum Pleiospilos Psammophora
Fouquieraceae	Idria	Idria
Liliaceae	Aloe Dracaena Gasteria Haworthia Yucca	Aloe Dracaena Gasteria Haworthia Yucca
Portulaceae	Anacampseros	Anacampseros

1   *Acanthocalycium violaceum* ($\times \frac{1}{2}$).   2   *Borzicactus sepium* ($\times \frac{1}{2}$).

3   *Ariocarpus fissuratus* ($\times \frac{1}{2}$).

4   *Astrophytum asterias* ($\times \frac{2}{3}$).

5   *A. capricorne* ($\times \frac{1}{2}$).

6   *A. myriostigma* var. *columnaris* ($\times \frac{1}{2}$).

7   *A. ornatum* (form) ($\times \frac{1}{2}$).

8   *Astrophytum capricorne* var. *senile* ($\frac{2}{3}$).

9   *Cephalocereus chrysacanthus* ($\times \frac{1}{2}$).

10   *Cephalocereus palmeri* ($\times \frac{1}{2}$).

11 *Cephalocereus senilis* ($\times\frac{1}{2}$).

12 *Cereus chalybaeus* ($\times\frac{1}{2}$).

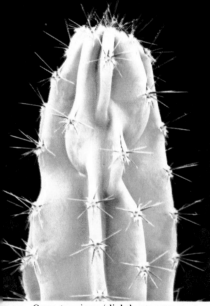

13 *Cereus peruvianus* (slightly monstrous form) ($\times\frac{1}{3}$).

14 *Cleistocactus jujuyensis* ($\times\frac{1}{2}$).

15   *Chamaecereus silvestri* ($\times \frac{1}{3}$).

16   *Chamaecereus silvestri* (hybrid) ($\times \frac{1}{3}$).

17   *Disocactus nelsonii* ($\times \frac{1}{2}$).

18   *Cleistocactus candelilla* ($\times \frac{2}{3}$).

19 *Copiapoa echinoidea* ($\times 1\frac{1}{2}$).

20 *Copiapoa humilis* (form) ($\times \frac{1}{2}$).

21 *Coryphantha clava* ($\times 1$).

22   *Coryphantha hesteri* ($\times \frac{4}{5}$).

23   *C. pallida* ($\times \frac{1}{2}$).

24   *C. radians* var. *minor* ($\times \frac{1}{2}$).

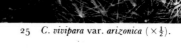

25   *C. vivipara* var. *arizonica* ($\times \frac{1}{2}$).

26　*Coryphantha nelliae* (× 1⅓).

27　*C. radians* (×1).

38   *Echinocereus dasyacanthus* (× 1).

39   *E. pectinatus* (× 1).

40   *Echinocereus triglochidiatus* var. *paucispinus* ($\times$ 1).

41   *E. triglochidiatus* var. *paucispinus* ($\times\frac{1}{3}$).

42   *Echinocactus grusonii* ($\times \frac{1}{10}$).

43   *Echinomastus macdowellii* ($\times \frac{2}{3}$).

44 *Echinopsis ancistrophora* ($\times \frac{1}{2}$).

45 *E. longispina* ($\times \frac{2}{3}$).

46 *Epiphyllum* (hybrid) ($\times \frac{1}{2}$).

47 *E.* (hybrid) ($\times \frac{1}{3}$).

48 *Epithelantha micromeris* var. *greggii* (×1).

49 *Espostoa lanata* (×½).

50 *Espostoa lanata* var. *mocupensis* (×½).

51 *Espostoa melanostele* (×½).

52   *Ferocactus acanthodes* ($\times \frac{1}{3}$).

53   *F. latispinus* ($\times \frac{3}{4}$).

54 *Gymnocalycium denudatum* (×½).

55 *Frailea asterioides* (×1).

56 *Gymnocalycium castellanosii* (×¾).

57   *Gymnocalycium bruchii* ($\times \frac{1}{3}$).

60   *G. netrelianum* ($\times \frac{1}{2}$).

58   *G. hossei* ($\times \frac{1}{2}$).

59   *G. mihanovichii* ($\times \frac{1}{2}$).

61   *G. platense* ($\times \frac{1}{3}$).

62 *Gymnocalycium schickendantzii* (× ½).

63 *G. prolifer* (× 1).

64 *G. saglionis* (× ½).

65 *G. stuckertii* (× ½).

66 *Haageocereus divaricatispinus* (× ¾).   67 *Haageocereus olowinskianus* (× ¾).

68 *Harrisia jusbertii* (× ⅓).

69　*Harrisia bonplandii* and *H. martinii* ($\times \frac{1}{2}$).

70　*Hatiora salicornioides* ($\times 1\frac{1}{2}$).

71   *Homalocephala texensis* ($\times \frac{1}{4}$).

74   *L. minuta* ($\times 1$).

72   *Lobivia haageana* ($\times \frac{1}{3}$).

73   *L. huascha* var. *rubra* ($\times \frac{3}{4}$).

75   *L. drijveriana* ($\times \frac{1}{2}$).

76 *Lobivia famatimensis* var. *aurantiaca* ($\times \frac{2}{3}$).

77 *L. hertrichiana* ($\times \frac{2}{3}$).

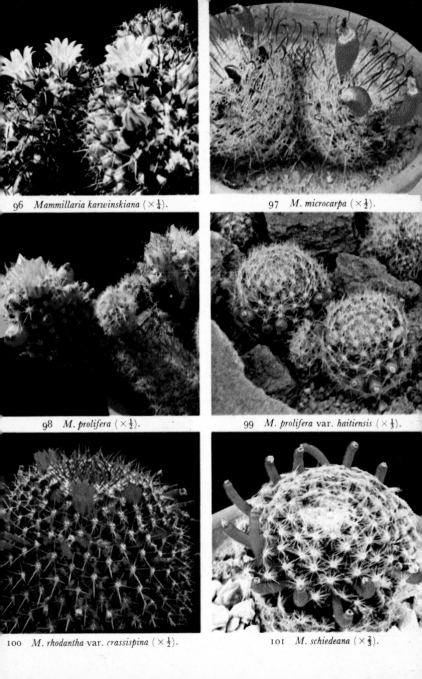

96 *Mammillaria karwinskiana* ($\times \frac{1}{4}$).

97 *M. microcarpa* ($\times \frac{1}{2}$).

98 *M. prolifera* ($\times \frac{1}{2}$).

99 *M. prolifera* var. *haitiensis* ($\times \frac{1}{3}$).

100 *M. rhodantha* var. *crassispina* ($\times \frac{1}{2}$).

101 *M. schiedeana* ($\times \frac{2}{3}$).

102 *Mammillaria scheidweileriana* ($\times \frac{1}{3}$).

103 *M. surculosa* ($\times \frac{1}{2}$).

105 *M. solisii* ($\times \frac{1}{2}$).

104 *M. wildii* ($\times \frac{1}{3}$).

106 *M. elongata* var. *stella-au*
($\times \frac{2}{3}$).

107 *Mammillaria sphacelata* (× 1).

108 *M. spinosissima* (× 1).

109   *Mammillaria tolimensis* ($\times \frac{1}{3}$).

110   *M. zeilmanniana* ($\times$ 1).

111 *Monvillea cavendishii* ($\times \frac{2}{3}$).

112 *M. spegazzinii* ($\times \frac{2}{3}$).

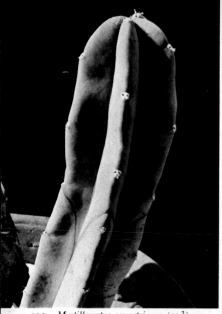

113 *Myrtillocactus geometrizans* ($\times \frac{2}{3}$).

114 *Neolloydia conoidea* ($\times \frac{2}{3}$).

115 *Neoporteria chilensis* ($\times 1$).

116   *Neoporteria nidus* fa. *senilis* ($\times \frac{1}{2}$).   117   *N. rupicola* ($\times \frac{1}{2}$).

118   *N. wagenknechtii* ($\times \frac{2}{3}$).

119 *Neoporteria jussieui* ($\times \frac{1}{3}$).

120 *Notocactus graessneri* ($\times \frac{1}{2}$).

121 *Nopalxochia phyllanthoides* ($\times \frac{2}{3}$).

122 *N. leninghausii* ($\times \frac{1}{2}$).   123 *N. scopa* ($\times \frac{1}{2}$).

124 *Notocactus leninghausii* fa. *cristata* ($\times$ 1).

125 *Notocactus mammulosus* (× 2).

126 *N. mueller-melchersii* (× 1).

127   *Notocactus ottonis* ($\times \frac{1}{2}$).

128   *N. scopa* fa. *ruberrima* ($\times 1\frac{1}{2}$).

129   *Opuntia durangensis* ($\times \frac{1}{3}$).

130   *O compressa* var. *macrorhiza* ($\times \frac{1}{3}$).

131   *O. leptocarpa* ($\times \frac{1}{4}$).

132 *Opuntia* species ($\times \frac{1}{3}$).

133 *O. bergeriana* ($\times \frac{1}{4}$).

134 *O. microdasys* fa. *alba* ($\times \frac{1}{3}$).

135 *Opuntia pachypus* fa. *cristata* ($\times \frac{1}{2}$).   136 *O. vestita* ($\times \frac{1}{4}$).

137 *O. verschaffeltii* ($\times 1$).

138 *Opuntia dimorpha* ($\times \frac{2}{3}$).

139 *O. molinensis* ($\times 1$).

140 *O. paediophila* ($\times \frac{1}{3}$).

141 *O. platyacantha* ($\times \frac{1}{3}$).

142 *Oreocereus celsianus* ($\times$ 1).

143 *O. ritteri* ($\times\frac{1}{2}$).

144 *O. trollii* ($\times\frac{1}{2}$).

145 *Pachycereus pringlei* ($\times\frac{1}{3}$)

146 *Parodia catamarcensis* ($\times\frac{1}{2}$).

147 *Parodia comarapana* ($\times\frac{1}{2}$).

148 *Parodia* species ($\times\frac{2}{3}$).

149 *Parodia aureispina* (× 1).

150 *P. macrancistra* (× ½).

151 *P. scopaoides* (× ⅔).

152   *Parodia chrysacanthion* ($\times \frac{1}{2}$).

153   *P. mutabilis* ($\times \frac{1}{2}$).

154   *P. rigidispina* ($\times \frac{2}{3}$).

155   *P. penicellata* ($\times \frac{1}{3}$).

156   *P. setifer* ($\times \frac{1}{3}$).

157   *Pelecyphora valdeziana* (×3).

158 *Pereskia aculeata* ($\times \frac{1}{4}$).

159 *Pterocactus tuberosus* ($\times 1\frac{1}{2}$).

160 *Rebutia aureiflora* ($\times 1\frac{1}{2}$).

161 *R. costata* ($\times 2$).

162 *Rebutia haagei* ($\times \frac{1}{2}$).

163 *R. pygmaea* ($\times$ 1).

164   *Rebutia calliantha* var. *beryllioides* (×½).

165   *R. fiebrigii* (×⅓).

166   *R. minuscula* var. *grandiflora* (×½).

167   *R. calliantha* var. *krainziana* (×½).

168   *R. marsoneri* (fa.) (×⅓).

169   *R. pseudodeminuta* (×⅔).

170 *Rebutia deminuta* fa. *pseudominuscula* (×2).

171 *R. senilis* (×½).

172 *R. senilis* var. *lilacino-rosea* (×⅔).

173 *R. spegazziniana* var. *atroviridis* (×⅓).

174 *R. xanthocarpa* fa. *salmonea* (×½).

175 *R. minuscula* fa. *violaciflora* (×⅔).

176 *Schlumbergera gaertneri* ($\times \frac{2}{3}$).

177 *Selenicereus grandiflorus* ($\times \frac{1}{2}$).

178 *Stetsonia coryne* ($\times \frac{1}{3}$).

179 *Stenocactus multicostatus* ($\times \frac{1}{2}$).

180 *Soehrensia bruchii* ($\times \frac{1}{2}$).

181 *Thelocactus bicolor* (×½)   182 *Thelocactus bicolor* (×½).

183 *Stenocactus crispatus* (×4).

184 *Trichocereus candicans* var. *gladiatus* (× ⅓).

187 *Toumeya schwarzii* (× ⅔).

185 *Trichocereus chiloensis* (× ⅓).

186 *Trichocereus shaferi* (× ¼).

188 *Trichocereus spachianus* (form (× ⅓).

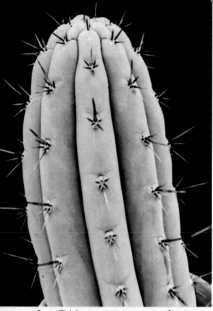

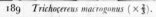

189   *Trichoçereus macrogonus* (× $\frac{2}{3}$).   190   *Wilcoxia poselgeri* (× $\frac{1}{2}$).

191   *Wilcoxia schmollii* (× 1).

192   *Trichocereus smrzianus* ($\times \frac{1}{2}$).

193   *Zygocactus truncatus* ($\times \frac{1}{4}$).

194 *Zygocactus truncatus* ($\times 1\frac{1}{2}$).

195 *Z. truncatus* (form).

196   *Aeonium arboreum* ($\times \frac{1}{3}$).

197   *A. arboreum* (close-up) ($\times 1$).

198 *Aeonium rubrolineatum* ($\times \frac{1}{4}$).

199 *Agave filifera* ($\times \frac{1}{10}$).

200 *Agave americana* var. *mediopicta* fa. *alba* ($\times \frac{1}{10}$).

201 *Agave americana* ($\times \frac{1}{10}$).

202　*Agave parrasana* ($\times \frac{1}{5}$).

203　*A. parviflora* ($\times \frac{1}{4}$).

204　*A. utahensis* var. *nevadensis* ($\times \frac{1}{3}$).

205　*A. victoria-reginae* ($\times \frac{1}{2}$).

206　*Aloe ciliaris* ($\times \frac{2}{3}$).

207　*A. deltoideodonta* ($\times \frac{1}{2}$).

208   *Aloe woolleyana* ($\times \frac{1}{2}$).   209   *A. humilis* (flower) ($\times 1$).   ·

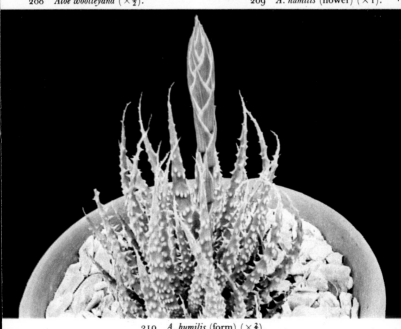

210   *A. humilis* (form) ($\times \frac{2}{3}$).

211 *Aloe striata* ($\times \frac{1}{8}$). 212 *A. peglerae* ($\times \frac{1}{2}$).

213 *A. striata* (flower) ($\times \frac{1}{4}$).

214 *Adromischus saxicola* (× 1).   215 *Adromischus trigynus* (× 1).

216 *Anacampseros rufescens* (× 4).

217 *Ceropegia barkleyi* ($\times 1\frac{1}{2}$).

218 *C. stapeliiformis* ($\times 1\frac{1}{2}$).

219   *Astridia hallii* (×2).

220   *Conophytum bicarinatum* (×1).

221   *Conophytum ernianum* (×1).   222   *C. proximum* (×1).

223   *C. fenestratum* (×1).

224 *Conophytum fraternum* ($\times 1\frac{1}{2}$).

225 *C. minutum* ($\times 1\frac{1}{2}$).

226   *Conophytum regale* ($\times 1\frac{1}{2}$).

227   *C. meyeri* ($\times 1\frac{1}{2}$).

228   *Cotyledon orbiculata* ($\times \frac{1}{2}$).   229   *C.* species ($\times \frac{2}{3}$).

230   *C. undulata* ($\times 1$).

231　*Caralluma europaea* (× 2).　232　*Crassula argentea* fa. *variegata* (× ⅔).

233　*Crassula argentea* (floweronly) (× 2).

234   *Crassula arborescens* ($\times \frac{1}{2}$).

235   *C. barbata* ($\times \frac{1}{2}$).

236   *C. ericoides* ($\times \frac{2}{3}$).

237   *C. pyramidalis* ($\times 1$).

238　*Crassula deceptrix* (× 1).

239　*C. volkensii* (× ⅔).

240 *Duvalia parviflora* (×2).

241 *Echeveria agavoides* (×1).

242 *Echeveria dactylifera* (× 1).

243 *E. pulvinata* (× ⅔).

Echeveria 'Ballerina' (McCabe hybrid) (×½).

245 *E. setosa* (×½).

246 *E. subrigida* × *shaviana* (×1).

247   *Euphorbia balsamifera* ($\times \frac{1}{3}$).   248   *E. flanaganii* fa. *cristata* ($\times 1$).

249   *E. breoni* ($\times 2$).

250　*E. milii* fa. *lutea* (× ⅔).

251　*E. razafinjohanii* (× ⅔).

252   *Euphorbia lophogona* ($\times \frac{1}{2}$).

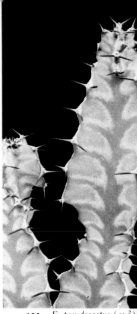

255   *E. pseudocactus* ($\times \frac{2}{3}$).

253   *E. anoplia* ($\times \frac{3}{4}$).

254   *E. neriifolia* fa. *cristata* ($\times \frac{1}{2}$).

256   *E. obesa* ($\times \frac{1}{2}$).

257　*Gasteria beckeri* ($\times \frac{1}{2}$).

258　*Glottiphyllum arrectum* ($\times \frac{2}{3}$).

259　*Gasteria liliputana* ($\times \frac{2}{3}$).

260　*Glottiphyllum oligocarpum* ($\times \frac{2}{3}$).

261　*Graptopetalum pachyphyllum* ($\times 1\frac{1}{4}$).

262　*Hereroa dyeri* ($\times 1$).

263 *Haworthia herrei* var. *depauperata* ($\times \frac{1}{2}$).

264 *H. reinwardtii* ($\times 1$).

265 *H. fasciata* ($\times \frac{3}{4}$).

266　*Imitaria muirii* ($\times \frac{2}{3}$).　　　　267　*Huernia keniensis* ($\times 1$).

268　*Huernia macrocarpa* ($\times 2$).

269 *Hoodia bainii* ($\times \frac{2}{3}$).

270 *Hoya carnosa* ($\times 1$).

271 *Kalanchoe blossfeldiana* (×⅓). 272 *K. fedtschenkoi* fa. *variegata* (×½).

273 *K. tomentosa* (×1).

274   *Kalanchoe beharensis* ($\times \frac{1}{2}$).

275   *Lampranthus peersii* ($\times$ 1).

276   *Lapidaria margaretae* ($\times 1\frac{1}{2}$).

277 *Lithops dorotheae* (× 1).

278 *L. salicola* (× $\frac{2}{3}$).

279 *L. olivacea* (× 1$\frac{1}{2}$).

280   *Monanthes polyphylla* ($\times$ 1).

281   *Orostachys japonicus* ($\times \frac{1}{2}$).

282   *Pleiospilos hilmari* ($\times \frac{3}{4}$).

283　*Ophthalmophyllum dinteri* (×1½).

284　*O. friedrichiae* (×2).

285　*O. praesectum* (×1½).

286 *Pachyveria* (hybrid) fa. *cristata* (× ⅓).

287 *Psammophora longifolia* (× ⅔).

288 *Sedum dasyphyllum* (× ½).

289 *Senecio stapeliaeformis* (× 1).

290 *Sedum humifusum* (× 2).

291 *Senecio stapeliaeformis* (× 1).

292　*Sansevieria hahnii* fa. *variegata* (× ½).

293　*Stapelia variegata* var. *marmorata* (× ⅔).

294　*Stultitia conjuncta* (× ⅔).

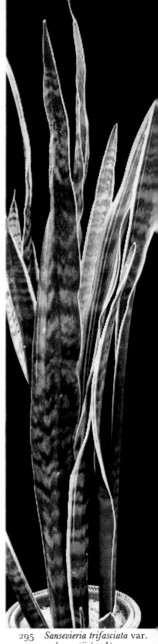

295　*Sansevieria trifasciata* var.
laurentii (× ¼).

296   *Stapelia nobilis* (×$\frac{1}{6}$).

297   *S. revoluta* (×$\frac{1}{2}$).

298 View from south end of main greenhouse in 'The Exotic Collection' ($\times \frac{1}{25}$).

299 *Parodia* staging view ($\times \frac{1}{5}$).

300   White stems of *Cleistocactus strausii*; in background multi-branched *Euphorbia grandider*
      ($\times \frac{1}{3}$).

301 *Browningia candelaris* ($\times \frac{1}{25}$).

302 *Ferocactus diguetti* ($\times \frac{1}{25}$).

303 *Haageocereus acranthus* (×$\frac{1}{12}$).

304 *Carnegiea gigantea* (×$\frac{1}{200}$).

305 *Haageocereus* species ($\times \frac{1}{12}$).

306 *Lemaireocereus thurberi* ($\times \frac{1}{60}$).

307 Large mounds of *Mammillaria parkinsonii* and flat prickly pads of *Opuntia* species ($\times \frac{1}{10}$).

308 *Melocactus bellavistensis* ($\times \frac{1}{6}$).

309 *Neoraimondia macrostibas* var. *roseiflora* ($\times \frac{1}{25}$).

310 *Opuntia floccosa* ($\times \frac{1}{12}$).

311 *O. versicolor* ($\times \frac{1}{2}$).

312 *Oreocereus hendriksenianus* ($\times \frac{1}{15}$).

313 *Weberbauerocereus* species ($\times \frac{1}{15}$).

314   *Aeonium palmense* ($\times\frac{1}{15}$).

315   *Aloe globuligemma* ($\times\frac{1}{15}$).

316   *Aloe dichotoma* ($\times \frac{1}{40}$).

317 *Dracaena cinnabari* ($\times \frac{1}{150}$).

318 *Euphorbia canariensis* ($\times \frac{1}{20}$).

319   *Euphorbia cooperi* ($\times \frac{1}{50}$).

320   *E. spiralis* ($\times \frac{1}{12}$).

321   *Euphorbia breoni* ($\times \frac{1}{15}$).

322 *Greenovia aurea* ($\times \frac{1}{10}$).

323 *Hoodia gordonii* ($\times \frac{1}{12}$).

324 Tall columns of *Idria columnaris* and large rosettes of *Agave* species ($\times \frac{1}{40}$).

325 *Yucca* species ($\times \frac{1}{20}$).

326   The tall thorny stems of *Pachypodium lamieri* growing amongst palm trees and other xerophytic shrubs ($\times \frac{1}{25}$).

12. DESCRIPTIONS OF SPECIES

The descriptions are listed in alphabetical order, according to the generic and specific names, but the true cacti are separated from the other succulents. The bracketed number following each name is the Plate Number. The Botanical Authority follows the name, and then the Group to which the Genus belongs is also given. The habitat of the genus concerned is stated and finally comes the Cultural Note; where a number of species are listed, unnecessary repetition is avoided by placing this Note after the final species. In the Cultural Note, you will find reference to the average soil mixture, which can be studied in more detail in Chapter 2. At the end of the section on Other Succulents, you will find details of two view pictures in the Exotic Collection, and these incorporate true cacti as well as other succulents.

NOTE: Var. = Variety and Fa. = Forma. These are used after the species' name where a lesser difference occurs. Fa. is used in the case of a rather minor difference.

CACTI

Acanthocalycium violaceum.
Bckbg. (1)
ECHINOPSIS GROUP

HABITAT: Argentina.
CULTURAL NOTE. Chiefly globular plants, occasionally becoming slightly columnar in age, strongly spined and free flowering. Require average soil mixture, prefer lightly shaded growing conditions and average water during all warm weather.

Ariocarpus fissuratus. *Sch.* (3)
ECHINOCACTUS GROUP

HABITAT: Mexico and Texas.
CULTURAL NOTE. Low growing plants, flowering in the autumn, but requiring 25% gritty sand in addition to the average soil mixture. Will grow and flower equally well in full sun or very lightly shaded position, but need less than average water.

Astrophytum asterias. *Zucc* (4)
Astrophytum capricorne. *Diet* (5)
Astrophytum capricorne *var.* **senile.**
Fric (8)
Astrophytum myriostigma *var.*
columnaris. *Tsuda* (6)
Astrophytum ornatum (short spined
form). *D.C.* (7)
These species belong to the
ECHINOCACTUS GROUP

HABITAT: Mexico, and one species in
Texas.
CULTURAL NOTE. Globular or sometimes
shortly columnar, all very free flowering
and often sweetly scented. Most species
may prefer 25% of gritty sand added to
the average soil mixture, but this will
depend to some extent on local climatic
conditions. Can take full sun, but a
lightly shaded position gives best results.
Slightly less than average water.

Borzicactus sepium. *B. & R.* (2)
CEREUS GROUP

HABITAT: Peru and Ecuador.
CULTURAL NOTE. An easy growing columnar species of free flowering habit,
which seems to apply to most of the
species within this genus. Average soil
mixture and watering, and a lightly
shaded position for best results.

Cephalocereus chrysacanthus.
B. & R. (9)
Cephalocereus palmeri. *B. & R.* (10
Cephalocereus senilis. *Pfeiff* (11)
These species belong to the
PILOCEREUS GROUP

HABITAT: Bolivia, Brazil and Mexico.
CULTURAL NOTE. Columnar growing
species, chiefly cultivated for their general appearance rather than their flowers.
Those growing them in the open air,
however, will find that once the plants

are large enough they will be rewarded
with many dozens of flowers on the upper
extremities. Require an average soil
mixture, with average water during
warm weather, although some care has
to be taken with the watering of *C.
senilis* when the plants are small. Will
grow well in full sun or light shade. The
first two species may be found here
under the Genus Pilocereus.

Cereus chalybaeus. *Orro* (12)
Cereus peruvianus. *Miller* (13)
(slightly monstrous form)
These species belong to the
CEREUS GROUP

HABITAT: Many parts of South America,
also West Indies.
CULTURAL NOTE. Very easy growing
columnar species, free branching when
large, and producing an abundance of
flowers on large plants. However, many
species are worthy of culture even as
small plants, and some of them can be
expected to flower when only about
3 ft (under 1 m) in height. Average soil
mixture and plenty of water in all warm
weather.

Chamaecereus silvestri. *B. & R.* (15)
Chamaecereus (hybrid). *Hort.* (16)
These species belong to the
CEREUS GROUP

HABITAT: Argentina.
CULTURAL NOTE: One of the most popular cacti, often the first species that many
people are given, as it is easily propagated
by cuttings and produces an abundance
of flowers. Average soil mixture, plenty
of water in all warm weather, but
partially shaded growing conditions are
important to prevent burning. Red
spider is very partial to this species, but
frequent overhead spraying will help to

deter this pest. There are numerous hybrids, many of them being the result of crossing with Lobivias. Most of the hybrids have rather robust stems and slightly larger flowers.

Cleistocactus candelilla. *Card* (18)
Cleistocactus jujuyensis. *Bckbg.* (14)
These species belong to the
CEREUS GROUP

HABITAT: Bolivia, Argentina and Paraguay.
CULTURAL NOTE. Very attractive columnar species, some of which are very free branching, mainly from the base, but most species will flower well even in a small greenhouse. Average soil mixture and watering; in winter a little water given occasionally can prevent tall specimens from bending over or even breaking off. Flowers appear mainly in spring or summer. *C. jujuyensis* is sometimes listed as a variety of *C. strausii*, which is shown in flower on Plate No. 300.

Copiapoa echinoidea. *B. & R.* (19)
Copiapoa humilis (longer spined form) *F.C.H.* (20)
These species belong to the
ECHINOCACTUS GROUP

HABITAT: Chile.
CULTURAL NOTE. Attractive globular species, which forms large clumps in age. Free flowering in summer; will grow and flower equally well in sun or under lightly shaded conditions. Average soil mixture and watering, but watering should be reduced or stopped in spells of cooler weather. Growers living in areas where long spells of cooler weather are experienced may find that the addition of 25% extra gritty sand is advantageous, and avoids any chance of over-watering.

Coryphantha clava. *Lem* (21)
Coryphantha hesteri. *Wright* (22)
Coryphantha nelliae. *Davis* (26)
Coryphantha pallida. *B. & R.* (23)
Coryphantha radians. *B. & R.* (27)
Coryphantha radians *var.* **minor.** *Hort* (24)
Coryphantha vivipara *var.* **arizonica.** *Marshll* (25)
These species belong to the
MAMMILLARIA GROUP

HABITAT: Southern United States and Mexico mainly.
CULTURAL NOTE. Coryphanthas vary from solitary globular species to other low growing varieties which form into large groups or mounds, whilst some are distinctly columnar in habit and can reach 1½ ft (0·45 m) in height. They are mostly easy growing, free flowering species, requiring average soil mixture and watering. However, they prefer a slightly shaded position to prevent burning, also plenty of overhead spraying to reduce the chances of red spider. Some species, such as *C. vivipara* and its varieties, can be termed hardy as they will withstand frost and snow if dry.

Disocactus macrantha. *K. & H.* (28)
Disocactus nelsonii. *Lindgr.* (17)
EPIPHYLLUM & PHYLLOCACTUS GROUP

HABITAT: Mexico.
CULTURAL NOTE. Two interesting epiphytic cacti, requiring the following treatment, i.e. 25% extra leafmould added to the soil mixture. They need plenty of water in all warm weather, some water in winter, and a partly shaded growing position. *Disocactus nelsonii* is also well known under the Genus Chiapasia.

Dolichothele sphaerica. *B. & R.* (29)
MAMMILLARIA GROUP

HABITAT: Texas and Mexico.
CULTURAL NOTE. Dwarf growing plants, all species possessing prominent tubercles with the spine cluster on the end, and some branch freely as this variety. Require an average soil mixture and watering, and lightly shaded growing conditions similar to those for the Coryphanthas.

Echinocactus grusonii. *Hild.* (42)
ECHINOCACTUS GROUP

HABITAT: Mexico.
CULTURAL NOTE. Our illustration shows some of our very large specimens of great age, but due to good treatment under glass in the Exotic Collection, they are almost without any scars. When grown in the open, they are not always so perfect. They are usually solitary, but in age some specimens do branch. Easy to cultivate, requiring an average soil mixture and watering. In winter they will stand cool dry conditions, but in damper climates a minimum of 45°F (7·2°C) may be safer to avoid unsightly marks appearing. This is particularly important when growing young plants of this specimen. Although not shown, flowers appear near the centre, and are of a similar colour to those of the spines.

Echinocereus dasyacanthus.
Engelm (38)
Echinocereus dasyacanthus *var.*
minor. *Engelm* (35)
Echinocereus enneacanthus.
Engelm (34)
Echinocereus fendleri. *Engelm* (30)
Echinocereus fitchii. *B. & R.* (36)
Echinocereus fitchii (another form).
B. & R. (37)
Echinocereus papillosus. *Link* (31)

Echinocereus pectinatus. *Scheid* (39)
Echinocereus reichenbachii. *Ters*
(32)
Echinocereus reichenbachii (another form). *Ters* (33)
Echinocereus triglochidiatus.
Engelm (40) *var.* **paucispinus**
Echinocereus triglochidiatus.
Engelm (41) *var.* **paucispinus**

These species belong to the
CEREUS GROUP

HABITAT: Southern United States and Mexico.
CULTURAL NOTE. We consider these amongst the finest of the larger flowered cacti, as the plants have many varying forms, beautiful coloured blooms, and unlike many cacti the flowers of most species last longer. It is not uncommon for a single flower to last 7–10 days and occasionally even longer. They are mostly low growing semi-columnar, either forming dense clumps, or irregular sprawling habit. Some, however, can remain solitary, particularly certain of the pectinate species. They require the average soil mixture and watering, although water may have to be reduced or stopped during any long spells of cool wet weather. Therefore in certain climatic conditions, some of the pectinate spined varieties can be more safely grown by the addition of another 25% gritty sand. Given lightly shaded conditions, they will grow and flower very freely. Many of the species will stand cool winter conditions if dry, and a few are winter hardy if dry.

Echinomastus macdowellii. *Rebut*
(43)
ECHINOCACTUS GROUP

HABITAT: Mexico and south-west United States.

CULTURAL NOTE. Normally small globular species of free flowering habit, requiring the addition of 25% gritty sand to the normal average soil mixture. Slightly less than average water at all times. These plants will grow in full sun or under lightly shaded conditions with equally good results.

Echinopsis ancistrophora. *Speg.* (44)
Echinopsis longispina. *Bckbg.* (45)
These species belong to the
ECHINOPSIS GROUP

HABITAT: Argentina and Bolivia.
CULTURAL NOTE. The commoner species, such as *E. multiplex* and *E. eyriesii* will probably be known to most growers, which is why we have illustrated two other fine species of this large flowered genus. They are mostly globular or shortly columnar species, requiring average soil and watering. Slightly shaded conditions give best results, and also prevent burning of the body of the plant. In winter most species will stand quite cool conditions with impunity. Mainly night flowering, but a few may last up to 36 hours or so.

Epiphyllum (hybrid). *Hort* (46)
(Often seen under *Phyllocactus*)
Epiphyllum (hybrid). *Hort* (47)
(Often seen under *Phyllocactus*)
These species belong to the
EPIPHYLLUM & PHYLLOCACTUS
GROUP

HABITAT: Mexico.
CULTURAL NOTE. These are two highly coloured hybrids, sometimes termed *Phyllocacti*. As a general rule the hybrids are more highly coloured than the true species, which is why we have illustrated the hybrids in preference to the others. They are epiphytic cacti, but are some-

what larger growing than such genera as Chiapasia, Disocactus, Zygocactus, etc., although requiring identical treatment. In other words they need an extra 25% leafmould with the average soil mixture, plenty of water in all warm weather, some water in winter, and a shaded growing position for best results. When plants are in bud a periodical feed with a liquid fertiliser can be beneficial.

Epithelantha micromeris *var.*
greggii. *Engelm* (48)
MAMMILLARIA GROUP

HABITAT: Mexico and Texas.
CULTURAL NOTE. Very dwarf, slow growing plants, which usually form into large clusters in age. They require an additional 25% gritty sand with the normal soil mixture, and less than average water at all times. Full sun or a lightly shaded position gives very good results. The flowers of this variety and the true species are pale pink, very small, and only last for a few hours. However, these flowers are produced quite freely in the summer, followed with luck by the red fruits which remain on the plants for a long time.

Espostoa lanata. *B. & R.* (49)
Espostoa lanata *var.* **mocupensis.**
Bckbg. (50)
Espostoa melanostele. *Vpl* (51)
These species belong to the
PILOCEREUS GROUP

HABITAT: Peru and Ecuador.
CULTURAL NOTE. Slow growing columnar plants, which usually branch from the base when fairly tall (6 ft or 1·8 m). Require average soil mixture, but slightly less than average water at all times, except in very warm weather when a little extra moisture can safely be given.

During long spells of dull cool weather, it may be necessary to stop watering. In areas of high humidity, it may be advisable to add an extra 25% gritty sand to the soil mixture. These plants are chiefly grown for their generally attractive appearance rather than their flowers; the blossoms produced by the larger plants are small and often almost hidden by surrounding hair.

Ferocactus acanthodes. *Lem* (52)
Ferocactus latispinus. *Haar* (53)
ECHINOCACTUS GROUP

HABITAT: Mexico and south-west United States.
CULTURAL NOTE. Globular to columnar, reaching a height of 9 ft (about 2·8 m) in a few species. Such large specimens will be up to 1½ ft (0·46 m) in diameter. However, most people grow this genus for the colourful spines rather than the flowers, as only a few species flower when relatively young. *F. latispinus* is one such species, which flowers when comparatively small. Require average soil mixture and watering, but watering may be stopped in spells of dull weather. A few species, such as *F. latispinus*, need a little extra warmth when young to avoid unsightly marks appearing, in much the same way as with young plants of *Echinocactus grusonii* (42).

Frailea asterioides. *Blossfld.* (55)
ECHINOCACTUS GROUP

HABITAT: Uruguay, Paraguay and Columbia.
CULTURAL NOTE. Very dwarf plants of hot weather. Partly shaded growing conditions are important for this species. The flowers are often cleistogamous. i.e. setting seed without ever opening, but in hot weather they will usually open for an hour or so in the afternoon.

Gymnocalycium bruchii. *Bckbg.* (57)
Gymnocalycium castellanosii. *Bckbg.* (56)
Gymnocalycium denudatum. *Pfeiff.* (54)
Gymnocalycium hossei. *Ber* (58)
Gymnocalycium mihanovichii. *B. & R.* (59)
Gymnocalycium netrelianum. *B. & R.* (60)
Gymnocalycium platense. *B. & R.* (61)
Gymnocalycium prolifer. *Bckbg.* (63)
Gymnocalycium saglionis. *B. & R.* (64)
Gymnocalycium schickendantzii. *B. & R.* (62)
Gymnocalycium stuckertii. *B. & R.* (65)
These species belong to the
ECHINOCACTUS GROUP

HABITAT: Argentina.
CULTURAL NOTE. Mostly globular species, some of which clump up in age, while others remain solitary. Very easy growing and flowering plants, requiring average soil mixture and watering, and a lightly shaded growing position. Most species will tolerate quite cool conditions in winter if dry. Some species possess a tap-root, so deeper pots are best.

Haageocereus divaricatispinus. *R. & B.* (66)
Haageocereus olowinskianus. *Bckbg.* (67)
These species belong to the
CEREUS GROUP

HABITAT: Peru.
CULTURAL NOTE. Densely spined columnar species, most of which have highly coloured spines. Plants usually branch

from the base, and are grown chiefly for plant form and colour rather than their flowers, which are invariably nocturnal and either white or pink. Mostly easy growing, requiring average soil mixture and watering, and only slight warmth is needed in winter for safe keeping. They grow well in full sun or under very light shade.

Harrisia bonplandii. *B. & R.* (69) (Sharply 4-angled stem with spine clusters)
Harrisia jusbertii. *B. & R.* (68)
Harrisia martinii. *B. & R.* (69)
These species belong to the
CEREUS GROUP

HABITAT: Argentina, Brazil, Paraguay and the East Indies.
CULTURAL NOTE. These are mostly tall clambering plants, which produce large white or greenish-white nocturnal flowers, followed by these brilliantly coloured fruits if they have been fertilised. Very easy growing, and none of the species can be termed slow growing, given the average soil mixture and plenty of water in all warm weather. Some authorities consider that the South American species, such as the three listed above, should be grouped under the genus name Eriocereus.

Hatiora salicornioides. *DC.* (70)
HATIORA & RHIPSALIS GROUP

HABITAT: Mexico.
CULTURAL NOTE. This is another epiphytic cactus, but one of much smaller habit and flower size. As with the previous plants of this kind, an additional 25% leafmould is needed, also plenty of water and overhead spraying, with some shade, to give the best growth and flower results.

Homalocephala texensis. *B. & R.* (71)
ECHINOCACTUS GROUP

HABITAT: Mexico, New Mexico and Texas.
CULTURAL NOTE. A most attractive globular cacti, closely related to the Genus Echinocactus. It is of easy culture given the average soil mixture and watering, and either full sun or lightly shaded growing conditions will ensure the best results. We prefer the latter, under glass.

Lobivia drijveriana. *Bckbg.* (75)
Lobivia famatimensis. *Speg* (81)
Lobivia famatimensis var. **aurantiaca.** *Bckbg.* (76)
Lobivia haageana. *Bckbg.* (72)
Lobivia hertrichiana. *Bckbg.* (77)
Lobivia huascha var. **rubra.** *Mshll.* (73)
Lobivia janseniana var. **leucantha.** *Bckbg.* (80)
Lobivia lateritia var. **schneideriana.** *Bckbg.* (78)
Lobivia minuta. *Ritter* (74)
Lobivia westii. *P.C.H.* (79)
Lobivia wrightiana. *Bckbg.* (82)
These species belong to the
ECHINOPSIS GROUP

HABITAT: Mainly Bolivia.
CULTURAL NOTE. These are mostly small globular or shortly columnar clustering plants, of very free flowering habit. Unfortunately the highly coloured flowers only last two days as a rule, but they are produced in reasonable profusion. Culture is easy, requiring average soil and watering, but a slightly shaded growing position gives best results. Overhead mist spraying is beneficial, and helps to reduce the chances of red spider. In winter most species will tolerate quite cool conditions if dry.

Mammillaria bocasana. *Poselg* (83)
Mammillaria bombycina. *Quehl.* (84)
Mammillaria candida. *Schd.* (85)
Mammillaria celsiana. *Lem* (86)
Mammillaria cowperae. *Shly.* (87)
Mammillaria crocidata. *Lem* (88)
Mammillaria elongata var. **stella-aurata.** *K. Sch.* (106)
Mammillaria gilensis. *Bdkr.* (89)
Mammillaria gracilis var. **pulchella.**
Salm-Dyck (90)
Mammillaria hahniana. *W. & B.*
(91)
Mammillaria karwinskiana. *Martius*
(96)
Mammillaria martinezii. *Teegel* (93)
Mammillaria mendelliana. *Werd*
(92)
Mammillaria microcarpa. *Engelm*
(97)
Mammillaria microheliopsis. *Werd*
(95)
Mammillaria pringlei. *K. Brand* (94)
Mammillaria prolifera. *Haw.* (98)
Mammillaria prolifera var.
haitiensis. *K.Sch.* (99)
Mammillaria rhodantha var.
crassipina. *K.Sch.* (100)
Mammillaria scheidweileriana.
Otto (102)
Mammilliara schiedeana. *Ehrb.* (101)
Mammillaria solisii. *Bdkr.* (105)
(form with few hooked spines)
Mammillaria sphacelata. *Mart* (107)
Mammillaria spinosissima. *Lem*
(108)
Mammillaria surculosa. *Bdkr.* (103)
Mammillaria tolimensis. *Craig* (109)
Mammillaria wildii. *Dietr* (104)
Mammillaria zeilmanniana. *Bdkr.*
(110)
These species belong to the
MAMMILLARIA GROUP
HABITAT: Mexico, south-west United
States, and parts of the West Indies.

CULTURAL NOTE. These are very popular
plants, due to their dwarf habit, whether
solitary or clustered, attractive spination
and the fine rings of flowers which most
species produce regularly each year.
Most species can be grown in our
average soil mixture, but the following
do require less than average water at all
times: *M. bombycina; M. martinezii;
M. microcarpa; M. schiedeana; M. solisii;*
and *M. tolimensis.* For those in damper
climates, the addition of 25% extra
gritty sand may be preferred for successful culture of these plants.

Monvillea cavendishii. *B. & R.* (111)
Monvillia spegazzinii. *B. & R.* (112)
These species belong to the
CEREUS GROUP

HABITAT: Argentina, Brazil, Paraguay
and Peru.
CULTURAL NOTE. Mainly columnar
plants, tending to require some support
after reaching 3 ft (just under 1 m) in
height. They require average soil mixture and watering, and do well under
light shade, when good growth and a
profusion of flowers can be expected.
In winter, however, a minimum of 45°F
(8°C) is safer for most species, unlike
many others in this Cereus group which
will stand much lower winter temperatures.

Myrtillocactus geometrizans.
Console (113)
CEREUS GROUP

HABITAT: Guatemala and Mexico.
CULTURAL NOTE. Columnar branching
species, but quite sturdy and not usually
requiring support in any way, unlike the
preceding species. Culture the same as
for Monvillea, with average soil and
water and the same wintering temperature.

Neolloydia conoidea. *D.C.* (114)
MAMMILLARIA GROUP

HABITAT: Texas and Mexico.
CULTURAL NOTE. Dwarf growing plants, either solitary or clustering depending on the species, and some becoming shortly columnar. Mostly require average soil and watering, although the latter can be reduced during long spells of cool weather. Free flowering plants which will do well either in full sun or under light shade.

Neoporteria chilensis. *B. & R.* (115)
(also known under *Neochilenia*)
Neoporteria jussieui. *B. & R.* (119)
Neoporteria nidus var. **senilis.** *Don & Rwly.* (120)
Neoporteria rupicola. *Don & Rwly.* (116)
(also known under *Horridocactus*)
Neoporteria wagenknechtii. *Rith* (118)
(also known under *Horridocactus*)
These species belong to the
ECHINOCACTUS GROUP

HABITAT: Chile.
CULTURAL NOTE. Globular plants in the main, either solitary or clustering, and mostly very free flowering in summer. Most species will do well in our average soil mixture, in full sun or under partial shade, with slightly less than average water. In winter they will tolerate quite cool conditions.

Nopalxochia phyllanthoides. *B. & R.* (121)
EPIPHYLLUM & PHYLLOCACTUS GROUP

HABITAT: Mexico.
CULTURAL NOTE. Another epiphytic cactus, which tends to flower very freely along the edges of the stems. Requires the extra 25% leafmould with the average soil mixture, plenty of water and some shade for good growth and profuse flowering. In winter some water will also be needed.

Notocactus graessneri. *Ber.* (120)
Notocactus leninghausii. *Ber.* (122)
Notocactus leninghausii *fa.* **cristata.** *Hort.* (124)
Notocactus mammulosus. *Ber.* (125)
Notocactus mueller-melchersii. *Fric or Bckbg.* (126)
Notocactus ottonis. *Ber.* (127)
Notocactus scopa. *Ber.* (123)
Notocactus scopa *fa.* **ruberrima.** *Hort.* (128)
These species belong to the
ECHINOCACTUS GROUP

HABITAT: Argentina, Brazil and Paraguay.
CULTURAL NOTE. Small globular or short columnar growing plants, of easy growth and profuse flowering habit. These—like Rebutias—are amongst the easiest plants for beginners to cultivate, and flower without any trouble. Require average soil mixture and watering, and prefer a lightly shaded position. In winter quite cool conditions will do no harm if dry.

Opuntia bergeriana. *Weber* (133)
Opuntia compressa var. **macrorhiza.** *Benson* (130)
Opuntia durangensis. *B. & R.* (129)
Opuntia leptocarpa. *Mcksn.* (131)
Opuntia microdasys var. **alba.** *Hort.* (134)
Opuntia pachypus *fa.* **cristata.** *Hort.* (135)
Opuntia species. (132)
(related to *O. azurea* and *O. santa-rita*)
Opuntia verschaffeltii. *Cels* (137)
Opuntia vestita. *Salm-Dyck* (136)
Opuntia (Tephrocactus) dimorpha. *Forst* (138)

Opuntia (Tephrocactus) molinensis. *Speg.* (139)
Opuntia (Tephrocactus) paediophila. *Cast.* (140)
Opuntia (Tephrocactus) platyacantha. *Salm-Dyck* (141)
These species belong to the
OPUNTIA GROUP

HABITAT: Nearly all parts of North, Central and South America, also West Indies, whilst some species have become naturalised in parts of the 'Old World'.
CULTURAL NOTE. A large and very varied genus; some species are very dwarf in habit, whilst others grow to tree-like proportions. Culturally there is little difficulty with any of them, although some species such as *O. pachypus* are slow growing. Most species will grow in almost any soil, but we would always recommend our average soil mixture, with an additional 25% sand for any slow growing varieties. Will grow equally well in full sun or under light shade, and nearly all species require plenty of water in warm weather. Ideal plants for new collectors, who can thus gain much valuable experience before going on to cultivate the more difficult species. In winter, many will tolerate cool conditions, some even many degrees of frost. Only *O. microdasys* fa. *alba* can be termed slightly tender, and particularly prone to develop orange spots in a damp cold winter.

Oreocereus celsianus. *Ricc.* (142)
Oreocereus ritteri. *Bckbg.* (143)
Oreocereus trollii. *Kppr.* (144)
These species belong to the
PILOCEREUS GROUP

HABITAT: Bolivia, Northern Chile and Peru.
CULTURAL NOTE. Very popular plants, mainly grown because of their very hairy and spiny appearance. Most species need to be a fair age before they can flower, and the tubular blossoms range from rose to red. They are not difficult plants to cultivate, and will do well in our average soil mixture, but with slightly less than average water particularly in dull cooler weather. Will grow equally successfully in full sun or slight shade, but spine development is probably better when in full sun. In winter they will stand cool conditions if dry, which is not surprising in view of the fact that they come from the High Andes mountains.

Pachycereus pringlei. *B. & R.* (145)
CEREUS GROUP

HABITAT: Mexico.
CULTURAL NOTE. An easy growing columnar cactus, which branches when tall, but this is only likely to occur if you are growing them in the open air. An attractive spiny species, requiring average soil and watering, and a slightly warmed greenhouse in winter to keep the frost away. Plants can be disfigured if the temperature falls too near freezing during a wet winter.

Parodia aureispina. *Bckbg.* (149)
Parodia catamarcensis. *Bckbg.* (146)
Parodia chrysacanthion. *Speg* (152)
Parodia comarapana. *Card* (147)
Parodia species. 1 (48)
Parodia macrancistra. *Bckbg.* (150)
Parodia mutabilis. *Bckbg.* (153)
Parodia penicellata. *Fechs and V.d. Steeg* (155).
Parodia rigidispina. *Fric* (154)
Parodia scopaoides. *Bckbg.* (151)
Parodia setifer. *Bckbg.* (156)
ECHINOCACTUS GROUP

HABITAT: Argentina, Bolivia, Brazil and Paraguay.

CULTURAL NOTE. Most globular solitary plants, known for fine clusters of brilliantly coloured flowers which appear from the centres of the plants in the summer. The majority have hooked spines, but we have illustrated two straight spined species, *P. chrysacanthion* and *P. penicellata*. Will grow well in our general soil mixture with slightly less than average water, but for those in a damper climate, the addition of some extra sand to give better drainage may be advised. This applies whether they are being grown in the open air or under glass.

Pelecyphora valdeziana. *Moellèr.* (157)
MAMMILLARIA GROUP

HABITAT: Mexico.

CULTURAL NOTE. All the species within this genus (of which we only illustrate one) are very slow growing, dwarf plants. They are not particularly difficult to cultivate, and will flower well given conditions of full sun or light shade, using the general soil mixture with an additional 25% sand. Require less than average water at all times.

Pereskia aculeata. *Mill.* (158)
OPUNTIA GROUP

HABITAT: Mexico and West Indies.

CULTURAL NOTE. A large growing plant, reminiscent of a rose because of its thorny stems, but relatively large leaves. These plants should only be cultivated if it is possible to let them grow up to 9 ft (2·8 m) or more, so that they can flower. In a good season you can expect such a plant to produce anything between five and ten thousand blooms during Octo-ber. Will grow in most soils, and appreciates plenty of water in all warm weather. In winter a little warmth to avoid freezing is best, although our largest plant of *P. aculeata* is grown in a cold greenhouse.

Pterocactus tuberosus. *B. & R.* (159)
OPUNTIA GROUP

HABITAT: Argentina.

CULTURAL NOTE. Low growing plants, possessing many stems which come from large underground roots. Each stem flowers at its tip before branching, but a few fine flowers can be expected annually from a small plant. The addition of extra sand to the normal soil mixture is usually recommended, with slightly less than average water at all times. It will stand cool conditions in winter.

Rebutia aureiflora. *Bckbg.* (160)
(also known under *Mediolobivia*)

Rebutia calliantha *var.* **beryllioides.** *Bng & Don.* (164)

Rebutia calliantha *var.* **krainziana.** *Bng & Don.* (167)
(also known as *R. krainziana*)

Rebutia costata. *Werd.* (161)
(also known under *Mediolobivia* and *Digitorebutia*)

Rebutia deminuta *fa.* **pseudominuscula.** *Bng & Don.* (170)
(also known as *R. pseudominuscula* or under *Aylostera*)

Rebutia fiebrigii. *B. & R.* (165)
(also known under *Aylostera*)

Rebutia haagei. *F. & S.* (162)
(also known under *Mediolobivia* and *Digitorebutia*)

Rebutia marsoneri. (form) *Werd.* (168)

Rebutia minuscula *var.* **grandiflora.** *Mshll. & Bock.* (166)
(also known as *R. grandiflora*)

Rebutia minuscula *fa.* **violaciflora.**
Bng & Don. (175)
(also known as *R. violaciflora*)
Rebutia pseudodeminuta. *Bckbg.*
(169)
(also known under *Aylostera*)
Rebutia pygmaea. *B. & R.* (163)
(also known under *Mediolobivia* and
Digitorebutia)
Rebutia senilis. *Bckbg.* (171)
Rebutia senilis *var.* **lilacino-rosea.**
Bckbg. (172)
Rebutia spegazziniana *var.* **atroviri-dis.** *Bckbg.* (173)
(also known under *Aylostera*)
Rebutia xanthocarpa *fa.* **salmonea.**
Bckbg. (174)
These species belong to the
ECHINOPSIS GROUP

HABITAT: Argentina and Bolivia.
CULTURAL NOTE. Dwarf free flowering,
clustering plants, which should be
amongst the first to be acquired by
those new to the hobby. They are easy
to grow both from cuttings and from
seed. Require the average soil mixture
and watering, and need a partially
shaded position for best results. Full sun
treatment can easily burn these plants.
In winter they will tolerate quite cool
conditions if dry.

Schlumbergera gaertneri. *B. & R.*
(176)
EPIPHYLLUM AND PHYLLO-
CACTUS GROUP
HABITAT: Brazil.
CULTURAL NOTE. Another small growing
and very popular cactus, often known
as the 'Easter Cactus'. Requires the
usual extra 25% leafmould with the
normal soil mixture, plenty of water in
all warm weather and some in winter.
Prefers a partially shaded position, parti-

cularly during the hotter months of the
year. As the plants produce so many
flowers, it is advisable to feed them at
weekly or fortnightly intervals with a
liquid fertiliser. This should also be done
when flowering has finished, as the
plants can become very weakened after
producing so many blooms.

Selenicereus grandiflorus. *B. & R.*
(177)
CEREUS GROUP

HABITAT: From Texas southwards as far
as Argentina and the West Indies.
CULTURAL NOTE. One of the most popular species is illustrated, but numerous
other species are known, all worthy of
culture and very easy to grow. They
require the average soil mixture, with
plenty of water in all warm weather.
They need support for their long clambering stems, but as aerial roots are
produced, they will support themselves
if grown against a wall or a tree. In
winter a little warmth is needed, but
can be kept dry.

Soehrensia bruchii. *Bckbg.* (180)
(also known under *Lobivia*)
ECHINOPSIS GROUP

HABITAT: Argentina.
CULTURAL NOTE. This plant has been
listed under various genera including
Lobivia, to which many experts still
think it belongs. An attractive species,
easy to grow from seed, which usually
remains solitary, but—unlike most Lobivias—can grow into very large single
heads. Can be expected to flower in five
years or so from seed, particularly if
given free root-run. Does well in most
soils, as it is not a fussy species, and
needs plenty of water in all warm
weather. Grows equally successfully in

full sun or partial shade, but we prefer the latter for best results. In winter, if kept dry, it will stand quite cool conditions.

Stenocactus (Echinofossulocactus) crispatus. *D.C.* (183)
Stenocactus (Echinofossulocactus) multicostatus. *Hild.* (179)
These species belong to the
ECHINOCACTUS GROUP

HABITAT: Mexico.
CULTURAL NOTE. Mostly globular solitary plants, but some do cluster eventually. Noted for its wavy ribs, which easily distinguish most species within this genus from those in any other. Strictly speaking, the generic name Echinofossulocactus has priority and should be used, but amateurs are more familiar with the genus name Stenocactus. These plants will do well in our general soil mixture with slightly less than average water. Flowers can be expected in the spring with most species, and they grow well in full sun or under light shade.

Stetsonia coryne. *B. & R.* (178)
CEREUS GROUP

HABITAT: Argentina.
CULTURAL NOTE. Like so many of the Cerei, this eventually becomes a tree-like plant, but when grown from seed it will take many years to outgrow a small greenhouse, even if given ample root room. It is worthy of culture because of its fine spination, as flowers cannot be expected until the plants are very large. Requires the average soil mixture and watering, and in winter should be kept well clear of frost, as unsightly marks can soon appear, particularly under damp cold conditions.

Thelocactus bicolor. *B. & R.* (181)
Thelocactus bicolor (another form). *B. & R.* (182)
These species belong to the
ECHINOCACTUS GROUP

HABITAT: Texas and Mexico.
CULTURAL NOTE. A very popular species which usually remains solitary and never becomes large, but a single plant can produce these fine flowers at intervals over many months of the growing season. An easy grower which does well in our normal soil mixture, with average water except in dull weather, when this can be reduced. In winter, if dry, it will tolerate quite cool weather.

Toumeya schwarzii. *Bio & Mshll.* (187)
(also known under *Strombocactus* and *Turbinicarpus*)
ECHINOCACTUS GROUP

HABITAT: Mexico.
CULTURAL NOTE. A very dwarf, free flowering, group of plants, which require an additional 25% sand with the general soil mixture, and need less than average water at all times. Will grow and flower successfully in full sun or partial shade, although we prefer the latter.

Trichocereus candicans *var.* **gladiatus.** *K.Sch.* (184)
Trichocereus chiloensis. *B. & R.* (185)
Trichocereus macrogonus. *B. & R.* (189)
Trichocereus shaferi. *Bckbg.* (186)
Trichocereus smrzianus. *Bckbg.* (192)
Trichocereus spachianus (form). *Ricc.* (188)
These species belong to the
CEREUS GROUP

HABITAT: Most parts of South America.

CULTURAL NOTE. The majority of species become reasonably tall, but the illustrations are of some of the smaller growing kinds, whch can be expected to flower even in a small greenhouse. For those growing them out of doors, there is wide scope, as there will be no limit to height and ultimate flowering size. Most species are easy to cultivate and will flourish in most soils, requiring plenty of water in all warm weather. In winter, a few species such as *T. shaferi* need a little warmth to avoid unsightly marks appearing.

Wilcoxia poselgeri. *B. & R.* (190)
Wilcoxia schmollii. *Wngt.* (191)
CEREUS GROUP

HABITAT : Texas and Mexico.
CULTURAL NOTE. Relatively dwarf, slender-stemmed cacti, producing these fine flowers even on very small plants. They are tuberous-rooted, but despite this they still need plenty of water in all warm weather, and our average soil mixture suits them well. In winter it is possible that stems may wither, which can be

avoided by giving a little water. Partially shaded growing conditions seem to produce the best results from these plants.

Zygocactus truncatus. *V. Sch.* (193)
Zygocactus truncatus. *V.Sch.* (194)
Zygocactus truncatus (another form). *V.Sch.* (195)
These species belong to the
EPIPHYLLUM AND PHYLLO-
CACTUS GROUP

HABITAT : Brazil.
CULTURAL NOTE. Probably the best known epiphytic cactus, of which there are many forms. A species which can be made to flower almost any time during the winter, depending on the temperature in which they are kept. As with other plants of this type, they need the additional 25% leafmould, plenty of water and some shade during the hottest weather. Frequent overhead spraying is advantageous, also a periodic feed of liquid fertiliser when the plants are in bud.

As you will have read earlier, true cacti belong to the family of Cactaceae, whereas the term 'succulent' covers a large number of plants from very many different genera and families. In this section on 'Other Succulents', the descriptions are given as for the section on Cacti: the Botanical Authority follows the name, and the family to which the Genus belongs is also given.

OTHER SUCCULENTS

Adromischus saxicola. *Sm.* (214)
Adromischus trigynus *v. Poelln.* (215)
These species belong to the
FAMILY CRASSULACEAE

HABITAT: South and South-West Africa mainly.

CULTURAL NOTE. Very easy-growing dwarf plants, most species of which are noted for their beautifully marked leaves, somewhat similar to the patterning on many birds' eggs. A few species, however, have plain leaves with perhaps a coloured margin. Will grow in most soils, but do well in the general soil mixture, with average watering in all warm weather. In winter they will tolerate cool conditions if dry. These plants are easily propagated by leaves, and are usually grown for their form as the flowers are rather uninteresting.

Aeonium arboreum. *W. & B.* (196)
Aeonium arboreum. *W. & B.* (197)
Aeonium rubrolineatum. *Sv.* (198)
These species belong to the
FAMILY CRASSULACEAE

HABITAT: Canary Islands mainly.
CULTURAL NOTE. Only the taller, shrubby species are shown in these illustrations, but there are some species similar in form though dwarf in habit, and others which are usually solitary—either on tall stems, or in some cases almost stemless. They are all easy to cultivate,

capable of growing in any reasonable soil mixture. The normal growing season is October to March, but most of the species can be encouraged to grow during the other months, when the weather is warmer in the northern hemisphere. The Canary Islands (although in the northern hemisphere) have their rains in the March-October period. Whereas most varieties would be damaged by frost, these plants do not need high temperatures, and are in fact ideal for bedding out of doors during the period May to October even in climates such as ours.

Agave americana. *L.* (201)
Agave americana *var.* **mediopicta** *fa.* **alba.** *Hort.* (200)
Agave filifera. *Salm.* (199)
Agave parrasana. *Bgr.* (202)
Agave parviflora. *Torr* (203)
Agave utahensis *var.* **nevadensis.** *Engelm* (204)
Agave victoria-reginae. *T. Moore* (205)
These species belong to the
FAMILY AGAVACEAE (formerly AMARYLLIDACEAE)

HABITAT: Mainly Central America and the West Indies.
CULTURAL NOTE. Popular easy-growing stemless plants, some species remaining solitary, such as *Agave victoria-reginae*, and others branching. Most species

withstand frost very well, if dry, but there are a few (not illustrated) which do require a little winter warmth. As with the Aeoniums, Agaves are very tolerant plants and will grow in most soils. They will take plenty of water in all warm weather, and are also ideal for outside planting in summer in climates such as that of the United Kingdom. In fact, many species will do extremely well in a cold greenhouse, and some will survive out of doors in winter if they are given some protection from the rain. When the flowers are produced the rosette dies, so these plants are grown mainly for their appearance rather than for their flowers; though the branching species soon replace their dead rosettes. The plants take many years to reach flowering size.

Aloe ciliaris. *Haw.* (206)
Aloe deltoideodonta. *Bak.* (207)
Aloe humilis form. Mill. (209)
Aloe humilis form. Mill. (210)
Aloe peglerae. *Schoenl.* (212)
Aloe striata. *Haw.* (211)
Aloe striata. *Haw.* (213)
Aloe woolleyana. *Pole Evans.* (208)
These species belong to the
FAMILY LILIACEAE

HABITAT: Most parts of South, Central and East Africa, various islands off the African coast, also Arabia, etc.
CULTURAL NOTE. A very varied genus, as some plants can be very dwarf in habit such as *Aloe humilis* (209 and 210), while others are climbing species such as *Aloe ciliaris* (206). In addition, there are many tree-like varieties, which we show as habitat pictures (see Pl. 316). Although not illustrated in close-up, *Aloe ferox* is commonly grown even in small greenhouses, and it takes many years to out-grow a building of this size. Like the

Aeonium and Agaves, these plants are mostly tolerant and easy to cultivate, but the use of our average soil mixture is best for the smaller growing species. Will stand full sun, but the smaller varieties mainly prefer a lightly shaded position, and they all like plenty of water in warm weather. In winter many will stand cool conditions, but certain East African and Madagascan species need a minimum temperature nearer the 50°F (10°C) mark.

Anacampseros rufescens. *Sweet.*
(216)
FAMILY PORTULACEAE

HABITAT: Mainly South Africa, but a few species in East Africa and one in Australia.
CULTURAL NOTE. Mostly easy growing plants, always dwarf in habit, but the whiter scaled types, which are not illustrated, do need extra care. The green-leaved kinds, such as this one, will do very well in the general soil mixture with average watering. A slightly shaded position is preferred, otherwise these varieties can become too bronzed on the leaves, which spoils their growth. The more difficult species mentioned earlier, such as *A. papyracea*, are best grown by people who have been collecting for a few years and have acquired some experience; these specimens can also take more sun.

Astridia hallii. *L. Bol.* (219)
FAMILY FICOIDACEAE

HABITAT: Mainly South Africa.
CULTURAL NOTE. Dwarf shrubby plants of easy culture and fairly free flowering habit. They will grow well in the normal soil mixture, with average watering, and a lightly shaded position.

Caralluma europaea. *N. E. Br.* (231)
FAMILY ASCLEPIADACEAE

HABITAT: North Africa.
CULTURAL NOTE. One of the many genera in the Stapeliad group of plants which possess these leathery-like five-lobed flowers. This species is dwarf in habit, branching above and below ground. Grows well in our normal soil mixture, with average water in warm weather and none in winter if kept on the cool side; a little should be given occasionally if they are kept in a temperature nearer the 50°F (10°C) mark. A partly shaded growing position gives the best results.

Ceropegia barkleyi. *Hook J.* (217)
Ceropegia stapeliiformis. *Haw.* (218)
These species belong to the
FAMILY ASCLEPIADACEAE

HABITAT: South and East Africa, also the Canary Islands.
CULTURAL NOTE. Easy growing plants, mainly of trailing habit, with the exception of the Canarian species (not illustrated) which have erect stick-like stems. Most of the trailing varieties grow among bushes in order to obtain protection from the sun and support for their rather weak stems. They will do well in normal soil mixture, with average watering and a partly shaded growing position. They do not like being subjected to frost, and the East African species need a higher minimum winter temperature.

Conophytum bicarinatum. *L. Bol.* (220)
Conophytum ernianum. *L. & T.* (221)
Conophytum fenestratum. *Schwant.* (223)

Conophytum fraternum. *N.E.Br.* (224)
Conophytum meyeri. *N.E.Br.* (227)
Conophytum minutum. *N.E.Br.* (225)
Conophytum proximum. *L. Bol.* (222)
Conophytum regale. *Lavis.* (226)
These species belong to the
FAMILY FICOIDACEAE

HABITAT: South and south-west Africa.
CULTURAL NOTE. Often known as dwarf Mesembryanthemums, these plants require very similar cultural treatment to the Lithops mentioned on p.182. They do well in our average soil mixture plus an additional 25% gritty sand, and have a fairly well defined growing season so the watering needs to be more strictly controlled, as follows: Dry from November to March; mist spraying or very light watering from April to June; normal overhead watering from July to October. This applies to the northern hemisphere, so should be changed by approximately six months for the southern hemisphere. In the resting stage the plants shrivel up into the old heads (or leaves) and look quite dead; this is quite normal, however, and once watering recommences a new head or heads will appear.

Cotyledon orbiculata. *L.* (228)
Cotyledon undulata. *Haw.* (230)
Cotyledon species. (229)
These species belong to the
FAMILY CRASSULACEAE

HABITAT: South, south-west and East Africa mainly.
CULTURAL NOTE. Only the evergreen species are illustrated, since these are the most attractive plants. They are mostly dwarf and shrubby in habit, grown mainly for their appearance rather than their flowers, although in some cases the flowers are quite attractive as with the

unnamed variety shown (229). They do well in our average soil mixture, with plenty of water in all warm weather, and a partly shaded growing position gives the best results. They will stand cool winter conditions if dry.

Crassula arborescens. *Willd.* (234)
Crassula argentea. *L.J.* (233)
Crassula argentea *fa.* **variegata.** *Hort.* (232)
Crassula barbata. *Thbg.* (235)
Crassula deceptrix. *Schoenl.* (238)
Crassula ericoides. *Haw.* (236)
Crassula pyramidalis. *Thbg.* (237)
Crassula volkensii. *Engl.* (239)
These species belong to the
FAMILY CRASSULACEAE

HABITAT: South Africa, and a few in other parts of the world.
CULTURAL NOTE. The majority of species are very easy to cultivate, requiring no special attention. They like plenty of water and growing positions providing broken sunlight, particularly the smaller species such as *C. ericoides* (236). However, *C. barbata, C. deceptrix* and *C. pyramidalis* do prefer a rather more sandy growing mixture than the others, and slightly less than average water at all times. *C. arborescens* and *C. argentea* are taller, shrubby species which become almost tree-like with age, but it does take a fair time to reach this height. Most South African varieties winter well in cool conditions just above freezing, but *C. volkensii*—which is from East Africa—needs a minimum temperature of 45°F (8°C) for safety.

Duvalia parviflora. *N.E.Br.* (240)
FAMILY ASCLEPIADACEAE

HABITAT: Mainly South Africa, with a few in East Africa.

CULTURAL NOTE. Interesting dwarf plants, easy to cultivate in our normal soil mixture, with average watering in all warm weather. They prefer partial shade to prevent burning, and, being one of the Stapeliad group, will produce masses of small spoke-like flowers. In winter most species will tolerate fairly cool conditions, with a little water occasionally to prevent the stems from becoming too shrivelled.

Echeveria agavoides. *Lem.* (241)
(sometimes known as *Urbinia agavoides*)
Echeveria 'Ballerina' (McCabe hybrid). *Hort.* (244)
Echeveria dactylifera. *E. Walth.* (242)
Echeveria pulvinata. *Rose.* (243)
Echeveria setosa. *R. & P.* (245)
Echeveria subrigida × shaviana (hybrid = ×). *Hort.* (246)
These species belong to the
FAMILY CRASSULACEAE

HABITAT: Mainly Central America.
CULTURAL NOTE. Very popular succulents because of their varied forms, some being stemless and others growing tall, also the fine pastel colours shown on the leaves and flowers of so many species. The majority are very easy to grow, using our average soil mixture. To develop their attractive leaf colouring, plenty of water should be given in the first half of the growing season, then the quantity can be halved in order to produce the fine leaf colours. Partial shade is best early in the season, but very little shade is needed later on. In winter, most species stand cool conditions well, when the rosettes will be smaller through losing the lower leaves, which is quite normal.

Euphorbia anoplia. *Staff.* (253)
Euphorbia balsamifera. *Ait.* (247)

Euphorbia breoni. *L.N.* (249)
Euphorbia flanaganii *fa.* **cristata.**
Hort. (248)
Euphorbia lophogona. *Lam.* (252)
Euphorbia milii (splendens) *fa.*
lutea. *Hort.* (250)
Euphorbia neriifolia *fa.* **cristata.**
Hort. (254)
Euphorbia obesa. *Hook. J.* (256)
Euphorbia pseudocactus. *Bgr.* (255)
Euphorbia razafinjohanii. *U. & L.*
(251)

These species belong to the
FAMILY EUPHORBIACEAE

HABITAT: Most parts of Africa and neighbouring islands, India, Ceylon, and a few species in the Americas.
CULTURAL NOTE. This genus probably has the widest range of plant forms, not only varying from very dwarf to tremendous tree-like varieties, but also including thorny, shrubby species, such as (249) to (251) from Madagascar, which bear non-succulent leaves. Many other shrubby species have similar leaves but smooth stems, such as (247), contrasting with the very succulent stemmed varieties such as (253) and (255). Although not illustrated, there are some species with large swollen root systems, or tubers, and many other intermediate forms. With such a wide range, there must necessarily be some variation in cultural requirements. The non-succulent leaved species require an ample supply of water when in growth, while those with the very succulent stems and no leaves prefer our general soil mixture with the addition of some extra sand, and less than average water at all times. We find that most species, even those with very succulent stems, prefer a slightly shaded position for best results. In winter, the majority of the numerous South African varieties only require slight warmth to keep them clear of frosts, while the East African, Indian and Madagascan species prefer a minimum temperature nearer 50°F (10°C) for safe keeping. All these species, particularly those with non-succulent leaves, need a little water occasionally in winter.

SPECIAL NOTE: Plate 248 depicts *E. flanaganii fa. cristata*, which is often seen incorrectly named as *E. caput-Medusae fa. cristata*, a much larger, thicker-stemmed species.

Gasteria beckeri. *Schoenl.* (257)
Gasteria liliputana. *V. Poelln.* (259)
These species belong to the
FAMILY LILIACEAE

HABITAT: South Africa.
CULTURAL NOTE. Very popular plants because of their easy culture, but we illustrate two of the lesser known species which are just as simple to grow. The flower spikes are fairly long, usually bearing pinkish-red tubular blossoms. Will grow in most soils, needing plenty of water in all warm weather. In winter they will stand cool conditions or even light frost, so long as they are dry.

Glottiphyllum arrectum. *N.E.Br.*
(258)
Glottiphyllum oligocarpum. *L. Bol.*
(260)
These species belong to the
FAMILY FICOIDACEAE

HABITAT: South Africa.
CULTURAL NOTE. Another genus in the Mesembryanthemum group, very dwarf in habit and having very succulent leaves. Flowers are mostly yellow, but occasionally white. Although these plants

will take plenty of water, they can be grown much more closely to their natural state by using our normal soil mixture with extra sand, and giving less than average water at all times. Without this treatment, they tend to become rather large and unnatural in appearance. They are, however, easy to cultivate, and watering can normally be carried out from spring to autumn. In winter they will tolerate cool conditions if completely dry. Some shade at all times is also important.

Graptopetalum pachyphyllum.
Rose. (261)
FAMILY CRASSULACEAE

HABITAT: Mainly Mexico.
CULTURAL NOTE. Although there are larger and better known species, this dwarf variety is a real gem and should be in every collection. It is very easy to grow, not particularly fussy over soil requirements, and likes average watering from spring to autumn. In winter it will stand cool conditions if dry.

Haworthia fasciata. *Haw.* (265)
Haworthia herrei *var.* **depauperata.**
V. Poelln. (263)
Haworthia reinwardtii. *Haw.* (264)
These species belong to the
FAMILY FICOIDACEAE

HABITAT: South Africa.
CULTURAL NOTE. Another popular group of easy growing plants, requiring treatment identical to that for Gasterias. Will thrive in most soils, with average water at all normal growing times, and none in winter when they can tolerate cool conditions. A partially shaded growing position is very important, and in the greenhouse most species—including even the rarer *H. truncata* (not

illustrated)—do well under the staging. With an excess of sun, the many plants in this group can easily become too bronzed.

Hereroa dyeri. *L.Bol.* (262)
FAMILY FICOIDACEAE

HABITAT: South Africa mainly in Cape Province, but one or two species are to be found in the Orange Free State and in south-west Africa.
CULTURAL NOTE. It is an interesting low growing shrubby plant, noted for its easy culture and free flowering habit, these usually opening during the latter part of the day and into the evening. Although belonging to the Mesembryanthemum group it does not require any special soil, and most species are very tolerant and can grow in most types of soil. They like plenty of water in all warm weather and if kept dry in the winter, will stand very cool conditions.

Hoodia bainii. *Dyer.* (269)
FAMILY ASCLEPIADACEAE

HABITAT: Angola, south-west Africa, and Cape Province in South Africa.
CULTURAL NOTE. These plants are thorny stemmed, forming into quite large clumps as can be seen in the habitat photograph of *Hoodia gordonii* (323). However, they will do well with careful culture, using our general soil mixture plus an additional 25% gritty sand, and giving less than average water except in very warm weather, when normal amounts can be given. Will grow in full sun, but we have found that slightly broken light conditions give the best results under glass. In winter they are best kept at a minimum temperature of 45°F (8°C), and should be reasonably dry.

Hoya carnosa. *R.Br.* (270)
FAMILY ASCLEPIADACEAE
HABITAT: China, India, Malaya, etc.
CULTURAL NOTE. Lovely clambering plants with leathery leaves, which produce these fine clusters of waxy flowers. Will do well in our average soil mixture, with plenty of water in all warm weather and some in winter, when a minimum temperature of 45°F (8°C) is preferred. Will tolerate lower temperatures, in which case they need to be kept rather dry and some leaves may be lost. It is not good for these plants to grow in full sun, as the leaves may burn. Do not remove the peduncles (stalks from which the flowers are produced), as more flowers will grow from them each year, in addition to the formation of new stalks.

Huernia keniensis. *R. E. Fries.* (267)
Huernia macrocarpa. *Spreng.* (268)
These species belong to the
FAMILY ASCLEPIADACEAE
HABITAT: South and East Africa mainly.
CULTURAL NOTE. This is another genus in the Stapeliad group, mainly comprised of rather smaller, densely clustering plants, very suitable for those with limited space. They are easy to cultivate using our normal soil mixture, and given average watering in all warm weather. Partly broken sunlight will give the best results. If a minimum winter temperature of 45°F (8°C) is possible, and a little water is given occasionally to prevent shrivelling, these plants can be brought into flower much earlier.

Imitaria muirii. *N.E.Br.* (266)
(also known as *Gibbaeum nebrownii*)
FAMILY FICOIDACEAE
HABITAT: South Africa.

CULTURAL NOTE. A slow-growing plant, which will do well in the general soil mixture, requires average watering except in long spells of dull, cool weather, when only very little should be given. The main growing period is summer and autumn, so less water should be given in the spring, particularly at the onset of only slightly warmer conditions. It will stand full sun, but broken sunlight gives the best results.

Kalanchoe beharensis. *D. del C.* (274)
Kalanchoe blossfeldiana. *V. Poelln.* (271)
Kalanchoe fedtschenkoi *fa.*
variegata. *Hort.* (272)
Kalanchoe tomentosa. *Bak.* (273)
These species belong to the
FAMILY CRASSULACEAE
HABITAT: Most parts of tropical Africa, India and the Far East.
CULTURAL NOTE. Exceedingly easy growing plants, varying from dwarf species to others which can reach a height of 6 ft (1·8 m). Very popular, not only because of their attractive leaves and beautiful clusters of flowers, but because they will grow in most soils so long as they are given plenty of water in all warm weather. In winter a minimum temperature of 45°F (8°C) is safest, though some of the very tropical species prefer it to be even higher and will need occasional watering to prevent undue leaf fall. Will grow in sun or partial shade, but the latter usually gives best results under glass.

Lampranthus peersii. *N.E.Br.* (275)
FAMILY FICOIDACEAE
HABITAT: Mainly South Africa.
CULTURAL NOTE. Exceedingly dwarf shrubby plants, which produce an abun-

dance of beautifully coloured flowers in the summer. Some species are ideal for summer bedding even in climates such as we have in the United Kingdom. In the greenhouse, the best results are obtained by giving them a free root-run, so that they can clump properly. They are not very fussy over soil requirements, but appreciate plenty of water throughout the growing season. Most species will stand quite cool conditions in winter, if dry.

Lapidaria margaretae. *Dtr &*
Schwant. (276)
FAMILY FICOIDACEAE

HABITAT: South-west Africa.
CULTURAL NOTE. One of the less common species of stemless Mesembryanthemums, requiring the general soil mixture with an additional 25% gritty sand, and less than average water at all times from spring to autumn. Will grow well in full sun or partially broken sunlight, the latter being preferred when under glass. This species will stand cool winter conditions if dry.

Lithops dorotheae. *Nel.* (277)
Lithops olivacea. *L. Bol.* (279)
Lithops salicola. *L. Bol.* (278)
These species belong to the
FAMILY FICOIDACEAE

HABITAT: South and south-west Africa.
CULTURAL NOTE. These attractive dwarf 'Stone Plants' (as they are often termed), are very popular with collectors all over the world. Most species are easy to cultivate, whether in tropical countries or in a climate such as in the United Kingdom. The growing period is mainly in the summer and autumn, and they require the average soil mixture with the addition of 25% gritty sand. In the

spring they need very light overhead spraying or watering; during the summer and autumn slightly less than average should be given, and none at all in winter when they will stand cool conditions if dry. Will grow in full sun, as they do in the wild, but slightly broken sunlight seems to give best results under glass. As they are very small plants, one may be tempted to put them in equally small pots, but in fact it is far better to grow a number of them—even of different species—in one larger pot, or better still in a wide pan. Certain troughs normally used for the rarer alpine plants make ideal containers for Lithops.

Monanthes polyphylla. *Haw.* (280)
FAMILY CRASSULACEAE

HABITAT: Canary Islands.
CULTURAL NOTE. A dwarf group of plants, of which this species is probably the finest, forming into dense clusters or mounds very similar to many alpine Saxifrages. They are very easy to cultivate, and are capable of growing in most soils although a rather sandy mixture does suit them best. Being Canarian plants, their normal growing season is from October to March, but they can be encouraged to do so at other times of the year if they are given water and a reasonable temperature.

Ophthalmophyllum dinteri. *Sch. &*
Jacobs. (283)
Ophthalmophyllum friedrichiae.
Dtr. & Schwart. (284)
Ophthalmophyllum praesectum.
Schwart. (285)
These species belong to the
FAMILY FICOIDACEAE

HABITAT: South and south-west Africa.

CULTURAL NOTE. This genus is closely related to Conophytums and Lithops, both of which have been mentioned earlier, and they require treatment identical to that for Lithops. In other words, they thrive on a rather sandy growing mixture, very light watering in the spring, less than average water in the summer and autumn, and none in winter. Broken sunlight gives the best results, as some species have been found to burn in full sun. In winter they will stand very cool conditions if dry.

Orostachys japonicus. *Bgr.* (281)
FAMILY CRASSULACEAE

HABITAT: Korea, Japan, etc.
CULTURAL NOTE. Attractive dwarf growing plants, many of which can be termed alpine succulents, as they will withstand frost and snow if in a very well drained soil. If they are to be grown in this way, and therefore subjected to cold, a well drained soil mixture is essential. In the growing season they will take plenty of water, and prefer conditions of broken sunlight.

Pachyveria (hybrid) *fa.* **cristata.** *Hort.* (286)
FAMILY CRASSULACEAE

HABITAT: Horticultural origin.
CULTURAL NOTE. These attractive succulents are hybrids between the two genera Pachyphytum and Echeveria. There are a considerable number of them, and the illustration shows a very fine crested specimen. The plants are fairly dwarf and clustering in habit, with very attractive leaf colouring as might be expected from their parentage. They will grow in most soils and are exceedingly tolerant of maltreatment! In the growing period they like plenty of water, although the leaf colour can be improved by reducing the amount for the second half of the season. Will stand cool winter conditions if dry.

Pleiospilos hilmari. *L. Bol.* (282)
FAMILY FICOIDACEAE

HABITAT: South Africa mainly in Cape Province.
CULTURAL NOTE. This is one of the more choice species in this genus, although most species are very popular, being of dwarf habit and very free flowering, as with the majority of the stemless Mesembryanthemums. In common with others in this group the additional 25% gritty sand, and less than average water at all times from spring to autumn is advised. It grows well in full sun, although partially broken sunlight is preferred when under glass. In winter stands cool conditions if completely dry.

Psammophora longifolia. *L. Bol.* (287)
FAMILY FICOIDACEAE

HABITAT: South-west Africa.
CULTURAL NOTE. One of the lesser known stemless Mesembryanthemums, but very easy to cultivate despite its relatively slow growth. Will do well in our average soil mixture with plenty of water during the summer and autumn. In winter and early spring it should be kept dry, when it will withstand cool conditions.

Sansevieria hahnii *fa.* **variegata.** *Hort.* (292)
Sansevieria trifasciata *var.* **laurentii.** *N.E.Br.* (295)
These species belong to the FAMILY AGAVACEAE

HABITAT: Many parts of tropical Africa, India, Ceylon, etc.
CULTURAL NOTE. Certain species of this

genus, particularly the two illustrated, have long been popular house plants. They generally make slow progress when pot grown in the house, but given a free root-run as is possible elsewhere their growth rate increases considerably. A somewhat sandy growing mixture is best, with slightly less than average water, except in very hot weather when more can be given. These plants will stand hot conditions, but the leaves can easily be burned if they are not given any shade. In winter most species prefer a minimum temperature of 50°F (10°C) or even higher, when little or no water is safest.

Sedum dasyphyllum. *L.* (288)
Sedum humifusum. *Rose.* (290)
These species belong to the
FAMILY CRASSULACEAE

HABITAT: This genus is found in most parts of the world, from northerly areas such as Iceland, southwards to Peru and Bolivia.
CULTURAL NOTE. Needless to say, the cultural requirements of these species vary considerably with such a range of habitats. Many varieties are grown as alpine plants in the open ground, where frost and snow occurs, but in this case a well drained soil is essential. Tender varieties, such as *S. humifusum*, will grow well in our average soil mixture and only require a minimum of warmth in winter; but all species will take plenty of water in the growing season. *S. dasyphyllum* is generally considered as a borderline alpine succulent, because it will withstand outside treatment during an average winter in this country. A severe winter would kill it, but specimens will do exceedingly well in an unheated greenhouse in winter. Most

species will grow equally well in full sun or partial shade.

Senecio stapeliaeformis. *Rwly.* (289)
Senecio stapeliaeformis. *Rwly.* (291)
These species belong to the
FAMILY COMPOSITAE

HABITAT: East Africa.
CULTURAL NOTE. One of the more popular species within this genus, due to the very attractive flower which lasts a reasonable time. It will branch above and below ground, and requires a fairly well drained soil; the general mixture plus additional sand is very suitable. Will take average watering in all warm weather, but in winter should be kept dry and preferably in a temperature not less than 45°F (8°C).

Stapelia nobilis. *N.E.Br.* (296)
Stapelia revoluta. *Mass.* (297)
Stapelia variegata *var.* **marmorata.**
N.E.Br. (293)
These species belong to the
FAMILY ASCLEPIADACEAE

HABITAT: South and south-west Africa, also East Africa.
CULTURAL NOTE. A very interesting genus with flowers ranging in size from very small to as much as 18 in. (approx. 0·46 m) in diameter. *S. nobilis* (296) usually has flowers up to 14 in. (approx. 0·38 m) in diameter. It is a very varied genus, and many species such as those illustrated will grow well in our soil mixture, with average water in all warm weather. Many varieties will stand cool winter conditions if dry, but if it is possible to provide a minimum temperature of 45°F (8°C) and a little water occasionally during the winter, they will flower a month or so earlier. The flowering period is long, some species

continuing to bloom over a period of 6–8 months. They mainly prefer conditions of broken sunlight to prevent the stems from becoming too bronzed, which also reduces the growth rate.

Stultitia conjuncta. *Wh. & Sl.* (294)
FAMILY ASCLEPIADACEAE

HABITAT: Rhodesia and South Africa.
CULTURAL NOTE. Another attractive genus in the Stapeliad group, requiring very similar treatment to the Stapelias previously mentioned. However, in winter a minimum temperature of 50°F (10°C) is desirable, with occasional water to prevent the stems from becoming too shrivelled. They also prefer some shade for the best growth and flowering results.

Exotic Collection views:
(298) In this picture you are looking northwards in the main greenhouse. which measures some 130 ft (about 40 m) in length. In it, you can see the white stems of *Cleistocactus strausii*, and a large blue-green *Aloe speciosa* in the centre, while a large plant of *Hoya carnosa* is growing on the trellis. Also visible are certain Trichocerei, small Oreocerei and the variegated leaves of *Yucca aloifolia*.

(299) This shows a section of one of the greenhouse stagings, which holds part of our fine collection of Parodias. You will have seen close-ups of the genus earlier on, but this gives a better idea of their free flowering habit. The pot diameters vary between 4 and 5 in. (10–12.5 cms).

(300) This picture shows just the tops of our tallest specimens of *Cleistocactus strausii*, illustrating the contrast between the flower colour and the white stems. Also shown is the top of a multi-branched specimen of *Euphorbia grandidens*, about 12 ft (3.6 m) in height.

13. DESCRIPTIONS OF SPECIES IN HABITAT

As mentioned in the Introduction, we are including a series of 26 colour illustrations equally divided between true cacti and the other succulents, which depict these plants in their native habitats. Such fine colour photographs will be of interest to all readers, but particularly to those in tropical climates, where large plants can be left to grow to their full size without the severe pruning required in more temperate climates because of the limitations imposed by greenhouses. They will give a much better idea of the immense size which some genera can attain, and how you can plan your tropical garden!

CACTI

Browningia candelaris. *B. & R.* (301)
CEREUS GROUP

A most unusual species when fully mature, because of the weird formation of the branches. It is native to Peru; this particular plant was filmed near Arequipa. As with many Cerei, it blooms at night, having nearly white funnel-shaped flowers.

Ferocactus diguettii. *B. & R.* (302)
ECHINOCACTUS GROUP

The majority of collectors think of Ferocacti as being very spiny globular plants, forgetting that in age many species become columnar and can reach 9–12 ft (about 2·5 to 3·6 m) in height. This particular specimen is native to southern California (*Baja California*), where it was filmed on the Island of Santa Catalina.

Carnegiea gigantea. *B. & R.* (304)
CEREUS GROUP

A monotypic genus, that is to say one with only a single species, mainly native to Arizona. This particular scene was filmed in the Saguaro National Park, 'saguaro' being the common name for the plant. Large specimens vary in height between 30 and 45 ft (about 9–14 m), and are likely to be a hundred years or so in age. The off-white flowers are produced near the tops of the stems, and the 'saguaro' is the state flower of Arizona.

Haagoocereus acranthus. *Ward & Bckbg.* (303)
CEREUS GROUP

A very fine grouped specimen is shown in this illustration, photographed at about 4,000 ft up (approx. 1,230 m) in Peru. Stems rarely reach more than 3 ft (just under 1 m) in the wild state, but in cultivation where they are away from wind, etc., specimens can grow rather taller. The flowers are pink, borne near the tops of the stems.

Haageocereus species. *Bckbg.* (305)
CEREUS GROUP

An unidentified species (Rauh 58110) showing a fine specimen, which has obviously flowered profusely because of the remains of hair on the sides of the stems.

Lemaireocereus thurberi. *B. & R.* (306)
CEREUS GROUP

A very fine specimen filmed 46 miles north of Culiacan, Sinaloa, Mexico, whose size can easily be seen as compared with Gilbert Voss, who is standing beside it. Flowers are funnel shaped, appearing near the tops of the stems, and are pinkish-red in colour.

Mammillaria parkinsonii. *Ehrbg.* (307)
MAMMILLARIA GROUP

Opuntia species. *Miller.* (307)
OPUNTIA GROUP

This is almost a natural garden setting, showing on the left two very fine clusters of *Mammillaria parkinsonii*, together with a brown spined Opuntia. Even in cultivation one can expect to grow some fine clusters of many Mammillaria species, but patience is essential since specimens such as those illustrated here could easily be over fifty years old! This view was filmed near Cadareyta, Mexico.

Melocactus bellavistensis. *Rauh & Bckbg.* (308)
ECHINOCACTUS GROUP

Although Melocacti are not among the easiest plants to grow, since most species require a fairly high minimum winter temperature of at least 50°F (10°C), no book would be complete without an illustration of these interesting plants. The unusual growth on top of each is known as a cephalium, from which the flowers are borne. A plant cannot flower until it has reached maturity and the cephalium is formed; generally speaking, once the cephalium starts to grow the rest of the plant does not get much larger. Most Melocacti are best grown from seed or grown on by collectors from seedlings only a year or so old, as larger plants—particularly those with cephaliums already formed—are not easy to re-establish.

Neoraimondia macrostibas *var.* **roseiflora.** *Rauh & Bckbg.* (309)
CEREUS GROUP

This is another Peruvian species in the Cerei group. Some specimens can reach 15 ft (4·6 m) or more in height, although the plants shown are not so large. However, this will give you a very good idea of the arid country from which many cacti come, and in fact at a distance we consider this species to be quite similar in form to *Euphorbia canariensis*, particularly those specimens which can be seen growing in equally arid volcanic regions in the Canary Islands. The flowers are very small on Neoraimondias, and are pink on this particular variety as the name suggests.

Opuntia (Tephrocactus) floccosa. *Salm-Dyck.* (310)
OPUNTIA GROUP

Very few collectors are likely to see specimens of such size, since unlike most Opuntias, varieties such as these are relatively slow growing. This comes from about 12,000 ft (3,700 m) up in the high Andes of Peru, and one can see the dense covering of white hair which is protection against the cold and which—together with the fact that the plants are dry—enables such species to survive the snow which occurs at these altitudes.

Opuntia versicolor. *Engelm.* (311)
OPUNTIA GROUP

This is just part of a 12-ft (3·7-m) high tree, filmed in Arizona. It is, as you can see, very densely spined and the spines are sheathed and barbed. If one has the space available, this is a fine plant for landscaping and will be massed in flowers if left undisturbed; the blooms of this species can vary from those shown to red or even purple in colour.

Oreocereus hendriksenianus.
Bckbg. (312)
PILOCEREUS GROUP

Another attractive high altitude species, filmed here in habitat at about 11,000 ft (3,400 m) in Peru. It is a fairly variable species, as the hair can vary from yellow or red to pure white as shown here. It has a clustering habit, and should be grown by everyone, as it rarely reaches more than 1½ ft (about 50 cms).

Weberbauerocereus species. *Bckbg.* (313)
CEREUS GROUP

A genus of plants in the Cerei group, but one which is not accepted as belonging to that group by all authorities. All the species are worthy of culture even when small plants, as they have dense spines, yellow, red or white in colour. They are not slow growers, and given a free root-run will soon make fine specimens, as can be seen from this habitat view taken near Arequipa in Peru.

OTHER SUCCULENTS

Aeonium palmense. *Webb.* (314)
FAMILY CRASSULACEAE

This species is usually a freely clustering plant, but the particular habitat scene shown, filmed south of Santa Cruz on the island of La Palma in the Canary Islands, depicts a very highly coloured specimen. They are wonderful to mix in with cacti in a rockery, particularly if planted on their sides between the rocks. Two similar species, *A. canariense* and *A. virgineum*, are easily obtainable.

Aloe dichotoma. *L.J.* (316)
FAMILY LILIACEAE

One of the slower-growing tree Aloes native to the arid areas of Namaqualand and south-west Africa. The particular specimen photographed here was over 20 ft (more than 6 m) in height, and was growing on the fringe of the Namib Desert.

Aloe globuligemma. *Pole Evans.* (315)
FAMILY LILIACEAE

This is one of a number of stemless species of Aloes to be found in many parts of Africa, but is very distinct because of the unusual formation of the branches of the inflorescence. It is native to the northern Transvaal and Rhodesia. The flower spikes would be about 3 ft (just under 1 m) in height. The majority of Aloes are very easy to grow, provided that they are given a free root-run.

Dracaena cinnabari. *Balf. J.* (317)
FAMILY LILIACEAE

Dracaenas are borderline succulents, but worthy of culture, although it would take very many years to grow a specimen to the size of this particular species. Filmed on the Zikhon peaks in the Bagghiher Mountains on the island of Socotra, it was growing at a height of 3,800 ft (about 1,170 m). Although growing off the east coast of Africa, it is closely related to the better known *Dracaena draco*, which is native to the Canary Islands off the west coast of Africa. These tree members of the Liliaceae family should not be ignored by collectors whose space is not limited.

Euphorbia breoni. *L.H.* (321)
These species belong to the
FAMILY EUPHORBIACEAE

This Madagascan species is related to the better known *E. splendens (milii)* but has thicker stems bearing longer thorns and larger inflorescences. The flowers (bracts) are also bigger, and when in leaf these are blue-green in colour. A wonderful species of easy culture, requiring plenty of water at all times for best results in cultivation. A close-up of this species is shown on Plate 249.

Euphorbia canariensis. *L.* (318)
FAMILY EUPHORBIACEAE

A well-known species easily obtainable by collectors, and one which lends itself to garden landscaping. This particular specimen is not tall, a mere 3 ft (less than 1 m) in height, as it was growing out of rock and in an exposed position very near the sea. In better localities the species can reach 12 ft (3·7 m) or more in height, and the larger clustered

specimens may be 30–40 ft (between 9 and 12·3 m) across. This variety is found on all the islands in the Canaries, and this particular photograph was taken near Buenavista in Tenerife.

Euphorbia cooperi. *N.E.Br.* (319)
FAMILY EUPHORBIACEAE

A very attractive species which can reach a height of up to 30 ft (just over 9 m). The branches are winged in appearance, but in age form a solid tree-like stem. It is not a slow grower, and will do well given a free root-run, but prefers a minimum winter temperature near 50° F (10° C). This was filmed in the Rhodesian bushveld, and the rather yellowish tipped stem is caused by strong sunlight; normally it is bright green in less brilliant conditions.

Euphorbia spiralis. *Balf. G.* (320)
FAMILY EUPHORBIACEAE

A rather less-known species as yet, since it is native to the island of Socotra, and was actually filmed at Jabel Hawart. The photograph will show you, as with the view of *E. cooperi*, that many succulent plants and trees often grow in association with ordinary deciduous trees, and are not always to be found in arid bare places.

Greenovia aurea. *W. & B.* (322)
FAMILY CRASSULACEAE

Another attractive Canarian succulent, particularly when seen like this as a large mass on a cliff face. As with *Aeonium palmense*, it is an ideal plant for rock niches; however, unlike *A. palmense* this species from high altitudes in the Canary Islands can withstand quite a few degrees of frost if dry. The flower spike is about 1 ft (30 cms) long, bearing a number of bright yellow flowers on the upper part. This view was taken below Cruz de Tejeda on Gran Canaria at a height of 5,500 ft (nearly 1,700 m). On the island of Tenerife, it grows at even higher altitudes.

Hoodia gordonii. *Sweat.* (323)
FAMILY ASCLEPIADACEAE

This genus belongs to the Stapeliad group, but these are rather slow-growing clustering plants coming from the same locality as *Aloe dichotoma* in the more arid areas of south-west Africa. The flowers are variable in size and colour, and can be up to 5 in (about 13 cms) across.

Idria columnaris. *Kellog.* (324)
FAMILY FOUQUIERACEAE

An unusual succulent to say the least, and one which makes you imagine that the plants are the wrong way up! This comes from near El Rosario, Baja California, where these plants grow in a very rocky soil along with Agaves and other cactus and succulent plants. They are slow-growing, but must be included if you like to have something unusual in your collection or garden.

Pachypodium lamieri. *Drake.* (326)
FAMILY APOCYNACEAE

Pachypodiums can be dwarf or tree-like specimens; the majority of species are native to Madagascar. They are very unusual succulents, and not too difficult to grow in cultivation provided a minimum winter temperature of 50°F (10°C) is maintained. Only the few South African species will stand lower temperatures. They all have attractive flowers varying from white to pink or even yellow. This view gives you an excellent idea of the mixed vegetation in which some succulents can be found. These comments also apply to many of the true cacti.

Yucca species. (325)
FAMILY LILIACEAE

These are also—like the Dracaenas—borderline succulents, some species of which can certainly be termed tree-like in habit. Many varieties such as *Y. gloriosa, Y. recurvifolia, Y. filifera, Y. kaufmanniana* are even suitable for the outside garden in cold climates where snow and heavy frosts occur. This particularly attractive scene was filmed in Monument Valley, Arizona.

14. GROWING FROM SEED

Simplicity with the very minimum of cost is the keynote to this section written to enable the beginner with absolutely no previous experience to make a success at the first attempt.

The method we have set down has many advantages which the beginner will soon recognise. No expensive apparatus is needed, the seed-raising mixture can be prepared by anyone with very little trouble, and once the seeds have been sown they can, in most instances, be left until the following spring or early summer *before* being pricked-out into new positions.

One of the most common causes for losses which many a beginner may have experienced is that of being tempted to prick-out the tiny seedlings too soon. At this stage of their growth, they are very delicate to handle, having extremely fine roots, and no matter how carefully the moving is done, invariably some are lost after the first early move. This risk may be completely avoided if the instructions in this book are followed faithfully.

After you have had a little experience in growing cacti from seed, you will be better able to judge the merits of other ways. While the use of expensive propagating frames, etc., may be successful in the hands of experts, the beginner will do well to gain experience in growing cacti from seed—the easy way.

'IDEAL' SEED-RAISING CONDITIONS

An understanding of the 'ideal' conditions in which cactus seeds will germinate, and the young seedlings continue to thrive until they are strong enough to be regarded as healthy young plants, is a great aid to success.

Not only can one judge the growth and proper development of the newly germinated seedlings but, perhaps even more important, those conditions which favour diseases likely to attack them can be more easily avoided.

It is worth remembering that cactus seedlings are often very small indeed, slow of growth and thus easily 'swamped' by any faster-growing disease, fungus or algae.

The ideal seed-raising conditions are:

1. Correct warmth.
2. Correct moisture.
3. Correct shading from direct sunlight or even bright daylight.
4. A free air circulation around the seedlings at all times.
5. A suitable 'seed-raising' mixture.

Take them in this order.

1. The night temperature should not fall much below 65°–70°F (18·3°–21·1°C), but the daytime temperature may often reach 90°F (32·2°C) or even higher.
2. Once the seeds have been sown and they have had a very thorough soaking, they require to be kept just moist, but not soaking wet, and they must *not* be allowed to dry out completely.
3. They must be shaded from direct daylight and, in particular, direct sunlight, yet not in complete darkness once seedlings begin to appear. (Later instructions deal with this more fully.)
4. A circulation of air around the seed pans, but not with cooling draughts, is important. This condition favours the cacti, but is not very favourable to disease, fungus or moss.
5. A seed-raising mixture made from leafmould and sand is ideal for the purpose, and is dealt with later in this chapter.

If the above conditions can be provided, a little extra care taken in the preparation of the seed-raising mixture, and a suitable position found to house the seed pans or pots, with provision for shading them based on the sketches on pages 207–8, then, together with the few simple requirements as detailed, there should be little to prevent a successful start being made in raising some cacti from seed.

One very important point is worth emphasising. Do not be too hasty in sowing seeds too early in the year. Ideal conditions are best from April until perhaps late in July, when no artificial heating is needed to guard against night temperatures falling too low. Also, those ideal conditions do not apply between January and March, and no electrical or other form of heating fully compensates for the more natural growing conditions which prevail for most types of plant life with the coming of spring.

The following simple requirements are set out, from which those you find most easy to obtain, or already possess, can be selected for future use. It will be seen that the essentials are few in number, but we have suggested some items which could serve as alternatives, or are optional, and their use is for you to decide.

Seed Pans You will require a few seed pans with holes in the base for water to be drawn up to supply moisture to the seeds before and after germination. Small pots can be used equally well but it is best for these to be not more than 2 in. (5 cms) deep. Whether pots or pans are used, their diameter does not matter very much, a convenient size being between 2 in. (5 cms) and 4 in. (10 cms) across. One or more holes in the base are essential.

Leafmould A small amount of old and well-rotted leafmould should be kept on hand, the very dusty kind being most suitable.

Sand A good gritty sand, such as often used on bowling greens, is excellent for seed raising.

Cotton Wool A small amount of cotton wool should be handy for use as required.

Water Clean water, preferably some which has been boiled to kill any unwanted organisms, should be stored ready for use. Rain water or tap water so long it has been previously boiled.

Dishes One or two flat dishes or trays suitable for standing the seed pans or pots into for germination and growing. These should be either glazed pottery, enamel, or perhaps plastic such as a photographic developing dish, and of a size to suit your own planning.

Sieve You will require a small sieve of standard gauge around No. 24, obtainable from any shop which sells cooking utensils. The usual cooking sieve is often around gauge 18, which is rather too coarse. Quite a small one is all you need; you may already possess one. Nos. 22, 24 or 26 are most suitable for our purposes. If in doubt about the gauge place a ruler on the sieve and count the number of 'holes' to 1 in. (2·5 cms), if there are between 22 and 26 to the inch, this will suit your needs.

Tweezers A pair of tweezers ought to be regarded as an essential; they have many uses and are not expensive. The blunt type is most suited to handling flat seeds, etc.

Mist Spray   While not perhaps absolutely essential, it is very useful. In fact, any spray which can produce a 'fine 'mist' is suitable—even a scent spray.

OPTIONAL REQUIREMENTS

The making of some form of 'hinged' cover for shading throughout the germinating and first season of growth is optional. Some form of shade *must*, however, be given, and this can either be very simple or a more professional job. You could, for instance, use an old picture-frame (without glass), a piece of wood at the back on which to hinge the frame, and muslin or meat-cloth to pin on the frame. These are ideal as they admit air through the material, yet shade the seedlings.

A few blocks of wood are required for the frame to rest on when closed. You could equally well use larger pots; it is for you to decide.

A simple method is to have blocks or pots at the four corners, lay a cane or lath across and shorter canes or laths the opposite way on which one or two thicknesses of thin paper, perhaps tissue paper, are laid. This serves the same purpose, but has to be removed and put back each time you water or examine the seed pans. It is worth the trouble to have something hinged at the back, but do *not* close in all round; remember that air should flow freely around and over the opts. When the frame is closed down, 2 in. (5 cms) clearance is advised, but exact distance is not important.

The use of a small heater of some kind is optional, depending on whether you wish to start very early in April. A thermometer is not essential, but is of course useful; one which registers both maximum and minimum temperatures is the best.

The use of potassium hydroxquinoline sulphate is optional.

A small water-can with a curved spout is most useful when adding small amounts of water to the germinating tray or trays, but is not essential.

PREPARING THE SEED-RAISING MIXTURE

The preparation of this simple and well-tried seed-raising mixture is worth a little extra care. Bearing in mind that your seedlings may remain in this mixture for about one year before being pricked out into new positions, it is important to make sure that it is free from anything harmful to germination or likely to develop during the first few months after the seedlings are growing. If you follow these instructions carefully, we are sure you will be well rewarded with good results.

Using the sieve mentioned on p. 193, sift out of your leafmould a small amount of the dusty content. This fine leafmould dust and slightly larger particles which have *passed through* the sieve from the basis for the mixture to be used.

Sifting this leafmould is much easier and far more satisfactory with dry, or almost dry, materials, and the same applies to the sifting of the sand which will be mixed with the fine dusty leafmould.

Using the same sieve, sift vigorously a quantity of sand, but it is the gritty sand which *remains* in the sieve which is to be used. Any rather large particles, or what might well appear as tiny pebbles, can be picked off with your tweezers so that you have remaining a *coarse, gritty sand* from which the fine dust has been taken out. (*Note:* If too much of the fine dust is allowed to remain it very often sets solid after a time and causes damage to young seedlings by 'cramping'.)

Always remember—the seed-raising mixture is to be made up from the dusty leafmould which has *passed through* the sieve, but the gritty sand to be used is that which *remains* in the sieve.

You will require to sift a larger quantity of gritty sand than leafmould, as the approximate proportions to be mixed together are: one part of dusty leafmould to which is added eight or nine parts of gritty sand.

Having mixed sufficient for your estimated needs, one other simple precaution is well worth while.

Place a newly prepared mixture in a tin, open tray or similar container, bake it in an oven for about one hour, or heat it up over a gas-jet or other flame or heater until too hot to touch. Allow this to remain hot for about one hour, then allow to cool off when you can store it in a clean container for your future use.

The purpose of this 'baking' is to kill anything which might be harmful to young seedlings. This is also sufficient to kill weed seeds, which can so easily be found in either leafmould or sand. If young seedlings are germinating and a few odd weeds begin to grow among them, the rapidly growing weeds are difficult to remove without dislodging the cactus seedlings and much harm can be done.

Baking, or just heating up, the seed mixture is not absolutely essential, if you are sure there is nothing in the mixture likely to cause trouble. We personally find the extra trouble in heating it up over a lamp or similar heater almost negligible, and strongly recommend you do this for safety.

Preparing the seed pans or pots in the same way by heating is a good idea if using anything not absolutely new. Previously used pots, etc., may well contain something harmful to your seedlings, and the heating up does away with this risk. All that is necessary is a large tin (such as a half-size biscuit tin), place pots, pans, etc., inside, put the lid on tightly and place over a heater or in the oven until too hot to touch, leave for a short time to make sure everything is thoroughly heated through, and you have clean materials free from harmful insects, weeds, etc.

It will always be found easier to prepare the seed pans or pots ready for the actual sowing beforehand. Use dry seed-raising mixture as this runs freely into position and does not need any pressing down; in fact the use of dry or even dust-dry mixture, sand, etc., is far more satisfactory and should not be wetted or damped *until* seeds have been sown.

Seed pans of different depths can be used, but for the purpose of these instructions we are assuming you are to use small pots. We use many small pots for seed sowing as they are very convenient for small quantities of named seeds.

Place at the bottom of each pot a piece of cotton wool, large enough to more than cover the hole but not packed too tightly, which would prevent water soakage. Pull odd pieces or strands of this cotton wool through the hole underneath and spread them around the hole. (This will act in much the same way as a lamp-wick when the pot is placed in water.) Press down gently.

On the cotton wool inside the pot run in some sand, which may be slightly damp (press down also), to cover the bottom of pot and wool. (This sand does not need to be sifted; a little finer sand does not prevent water being drawn up.)

Now run into each pot your seed-raising mixture. Do *not* fill the pots to the rims; leave about one-third of an inch (just under 7 cms) free above the mixture. Tap each pot to level the mixture which, being dry, will run freely, and if it appears lower around the rim of the pot, run a little extra round the outer edges, as this will prevent seeds falling too near the outside of the pot.

If you use any pots or pans which do not have large holes at the base, it may not be possible to draw the cotton wool through to the underside, *but* it is still advisable to place a thin layer of the wool across the bottom inside. Do not forget to cover the wool with sand as this settles firmly down and will prevent sinkage of seed mixture later on, particularly when it is first soaked.

Note: To sow seeds and then see the soil begin to sink unevenly can be disastrous, so make sure all is well before actually sowing. Here again, a 'test pot' of mixture *without* any seeds is worth while. After you have done everything as described, stand the pot in a tray of water. Note if it soaks up properly *and* whether the soil sinks; if all is well, then you can assume the others have been made up correctly. If, however, sinkage is observed, this means that when making up the pots for use you should have pressed the cotton wool down more tightly, and pressed this again after placing the layer of sand on the bottom. Do not press the seed-raising mixture down, however.

According to the number of seeds to be sown, you will require at least one dish in which they will stand after sowing, under some form of shaded position—preferably something similar to that given in the drawings on pages 207–8. You will also require another dish or tray in which to stand each pot or pan as it is sown, while it takes up its first thorough soaking.

Have these trays or dishes ready, and your shaded position prepared for the newly sown pots, where they are to remain for several months.

Clean, previously boiled water, tweezers, etc., should also be handy, thus completing this part of your preparations.

One other item is advisable but not essential. It will often be found more convenient to make a list of the names of the various species of seeds you are sowing, but instead of having large name labels in each seed pot or pan, a small label with a simple code number is worth using. You have only to refer to your list for the name of each sowing as and when you need to do so.

Note: Insert the labels (whether small or large) into each pot while the mixture is still dry, then sprinkle extra gritty sand near each label to prevent any seeds falling into the deeper cavity which it makes.

SOWING THE SEED

Seed-sowing might well be divided into four methods, although they differ only slightly according to the seeds being sown. All seed pans or pots are prepared in the same way as previously explained— cotton wool at the base, then sand covering and finally seed-raising mixture run into each pot to within about one-third of an inch (just under 1 cm) of the top.

Cactus seeds, and those of other succulents, quite naturally vary in size. Some are small, or even like dust, while many are fairly large and very easy to pick up singly. We may reckon, therefore, for our purposes that the seeds to be sown may be either:

1. very small;
2. medium average;
3. fairly large or large.

Each type sown might require some slight extra care or difference in method, but growers need not be alarmed by this.

Many beginners make a start by sowing mixed seeds; these may be of various sizes, but will be dealt with later in this chapter. They can be sown with equal anticipation of good results.

To take the seed sizes in order:

1. *Very Small Seeds*

On top of the previously prepared seed pot in which the mixture awaits the seeds, first sprinkle a thin layer of *gritty sand* lightly over the whole surface area. Sprinkle the very small seeds as thinly as possible; most of them will fall in between the coarse sand. Place the pot in the tray or dish of water (not the place where they are to remain for germinating), and allow this to soak until very wet. To ensure the seeds have all fallen between the gritty sand, use your mist-spray to help wash any remaining on top down to where they can germinate and find support. Allow the pot to soak for a time, ten minutes or even longer, then stand it in the dish or tray where it is to remain, with others to follow.

2. *Medium-sized Seeds*

Scatter seeds thinly over the surface of the mixture, stand the pot in the tray of water and allow to soak thoroughly. The medium-sized seeds will be seen on the wet surface, and being damp will not roll around as they would do on a dry surface. With thumb and forefinger, sprinkle enough *gritty sand* over the whole area until the seeds are just hidden from view. Do *not* bury them deeply. They should be almost, but not quite, visible.

After thoroughly soaking, transfer to the dish along with the pot first sown.

3. *Large Seeds*

This section might be divided into two parts, one for those larger seeds which are rounded in shape, and the other for flattened seeds such as Stapeliads, which are best sown singly with tweezers.

Let us first deal with the large seeds of the rounded type, of which many can be found among cacti.

Scatter them as evenly as possible over the surface with tweezers, move any clusters apart so that they will have room after germinating, and then—with a *very dry finger*—lightly press these big seeds slightly into the seed mixture, *not deeply* but so that each is resting in a depression, where it will remain firmly. Now stand the pot in the water-dish for its first thorough soaking, allow ten minutes or so— longer at this stage does no harm. (This applies to the *first* soaking for any of the seeds.)

When thoroughly soaked, use thumb and forefinger to sprinkle some *gritty sand* over the whole area, but this time the seeds (being larger) should be covered to approximately their own depth. Then transfer to the germinating position along with those previously sown.

With larger seeds of the *flattened* type (Stapeliads in particular), these are best picked up individually with tweezers and pressed into the seed-mixture so that the top edges of the seeds are just below the surface. You can easily space them by this method, and after sowing soak thoroughly as with previously pots, sprinkle gritty sand over the whole surface—as with the others—but use slightly more to give added covering. Do not bury them too deeply, but up to about a quarter of an inch (just óver $\frac{1}{2}$ cm) below the surface is about right for this type of seed.

You can judge the right depth by the size of seed; the very large ones can go deeper than those not so large. If the gritty sand is always used for this final covering, seedlings come through quite well because this will not clog as finer sand might do.

Sown in this way, they cannot lie flat on the surface as they would if scattered in the same way as the small seeds. Stapeliads in particular should be 'edge-down' and this can *only* be done with tweezers or perhaps (rather more clumsily) with thumb and forefinger.

SOWING MIXTURE SEEDS

As previously mentioned, mixed cacti, or mixtures containing both true cacti and other succulents, will contain a variety of sizes. It is not possible to give individual sowing treatment to each size. Large seeds or flattened kinds can probably be pressed into the mixture after they have previously been scattered on the surface, but medium and small seeds cannot be so treated.

The best and easiest way to sow mixed seeds is this: Have available some object with a smooth surface, a piece of label material, a piece of glossy card, or perhaps the bottom of a china pot or cup. Use something like this to press gently *all* seeds evenly down to soil level; if you use anything not so smooth, it may pick up some seeds instead of pressing them down.

When they are all evenly pressed into the seed mixture, very lightly cover with gritty sand, but not too thickly or the small seeds will not have sufficient room to extend the newly germinated seedling above this top layer.

Make sure the pot of mixed seeds is thoroughly soaked, as with all those previously sown, before transferring to the more permanent position.

To summarise briefly the procedure for sowing the seeds: Make sure your pots are filled carefully with mixture as recommended, leaving about one-third of an inch (just under 1 cm) free above the mixture to the rim of the pot. Sow seeds as well spaced and evenly as you can, covering with gritty sand according to the size of seed set out in the foregoing instructions.

Thoroughly soak each pot in a tray or dish of water put near by for this purpose—the first watering—before finally sprinkling the gritty covering over the newly damped seeds. (Except with the very small seeds which do not have to be specially covered.)

AN EXTRA PRECAUTION

A mild solution of potassium hydroxyquinoline sulphate, made from this fine powder and dissolved in the water in which the first soaking of the seed pans is made, does greatly guard against newly germinated seedlings damping-off. If all your pots, soil, sand, etc., are perfectly clean to begin with, this extra precaution may not be necessary, particularly as you will be growing from seed in the way we have recommended.

Damping-off of seedlings is more often found troublesome when seeds are within some totally enclosed frame, but where you have a free air circulation, the risk is much less. However, for those of you who wish to be especially thorough I suggest you obtain a small amount of this from your chemist for use now or at some later date. One-eighth to one-quarter of an ounce will probably last for a season or two and is inexpensive.

If you decide to use this when sowing your seeds, its strength does not require to be measured with absolute accuracy. Approximately the amount of this fine yellow powder which will remain piled up on a sixpence should be added to one pint (just over half litre) of water, and the solution will be about the right strength.

TREATMENT DURING THE FIRST SEASON

We have so far made no mention of the use of any form of artificial heating for maintaining a reasonable night temperature. If you do not sow your seeds until well into April or later, there should be little if any need for additional warmth at night.

You may, however, have a small heater of some kind. The ordinary oil-heater is quite suitable—we regularly use this type ourselves; or perhaps you may have the type of heater used for preventing freezing under the bonnet of a car. Anything of this kind may be useful but is not essential.

Should you feel you wish to have this available, arrange your seed-raising position in such a way as to allow for the small heater to be *under* the staging, perhaps on the floor. On particularly cold nights for the time of year, quite a modest amount of warmth, rising through the staging and passing freely around the position where the seed pots are kept, will give protection, and that extra warmth is helpful to quick, regular germination.

Seed sown after April will rarely need this added heat, and if you do not sow too early in the year germination is usually more rapid and more complete. Frankly, we think you gain little or nothing by sowing too early in the year, and have proved this many times by trials with both methods.

TREATMENT AFTER SOWING

Your seeds have been sown, they have had their first thorough soaking in the dish or tray *outside* their present position, and are now housed for some months to germinate and thrive.

Into the tray in which they are now standing, add a very small amount of water. Once this has been soaked up, try to keep the bottom of the tray *just moist*, a very thin layer of water only present always—not any measurable depth.

Once or twice each day, check this tray for dampness and try to avoid it completely drying out for long, if at all.

On very warm days, rather more frequent checking could be made if you are able to do so, but *never* fill it up. The seeds germinating do not require more than mere moisture, and too much water causes loss.

Make sure the tray is level so that water does not form to one side, leaving a dry patch somewhere else. If you can keep the surface of the tray *just damp* with any very slight actual water visible, your cotton-wool 'wicks' to each pot will draw the moisture needed to maintain correct germinating conditions.

Until some seedlings are visible, you can keep the shading rather darker if you wish—a sheet of newspaper on top, for instance—but once you begin to see the tiny plants appearing, keep shaded but not too dark. This can be judged by the colour of the seedlings, about which we will say more later.

GERMINATION

After sowing, you should see some results from between three days to perhaps three weeks. Some seeds germinate within twenty-four hours, others may take several weeks, while a few species can take much longer, but beginners will not be concerned with those.

If you begin to see some results from your labour in one to three weeks, you can regard this as satisfactory. All seeds will not germinate at the same time, and quite often occasional seedlings will appear for many weeks after sowing. Weather and temperature play a big part in all forms of growing, whether you use artificial warmth or not.

During the time the various pots are showing new germination, keep the moisture as constant as you can in the way described. Extra care in this for the first two or three months pays handsomely; watering rules become easier after three months.

If you find the odd pot or pan *not* taking up its water properly, the surface being dry and failing to moisten after extra moisture has been given, lift this pot from its tray and stand it in some deeper water, watch carefully and the moment you see any part of the top surface becoming wet, remove it at once to avoid over-soaking. By the time it has been put back into position it will be moist over the whole surface. Always use tepid or slightly warmed water in which to stand pots.

Newly germinated cactus seedlings are little more than tiny blobs of jelly; they are delicate and the first root is very small, perhaps only one thin hair-like projection seeking a way to fix into the soil.

For the first three months after sowing, water only by the method described—namely by keeping the tray or dish in which they stand *just moist*. Do *not* use your mist-spray over the top once the seedlings have begun to germinate. As they are very delicate at this stage, it may wash them out, dislodge the very small roots and thus cause loss.

After three months, watering becomes simpler. Overhead mist spraying can be given on warm days in addition to the moisture in the tray being kept as previously described.

Now a word or two about the right amount of shading. Small cactus seedlings are usually pale green to pinkish in colour (varying slightly with the species of course), but if you notice all seedlings more than about two months old becoming very red or bronzed, this indicates that they are receiving too much light, so give more shading. If they all appear very pale green or whitish, or seem to be trying to lean and stretch towards the light, then they are too heavily shaded. Adjust accordingly, and after a few days their colour will change to the more natural shades indicated.

Some of the seedlings will now be showing small spines. Roots will be developing fast at this stage and during summer weather the shading can be removed for an hour or so in the early morning, then they can be watered overhead with a water-can fitted with a fine rose; use the rose *upwards* to allow water to fall lightly over the whole area. After watering, replace the shading.

Water may not now be necessary every day, certainly not for all the plants. Some pots at least can be taken from the tray and watered overhead as described, perhaps two or three times each week. Growth in size may not be rapid with many of them, but if spines are developing well the plants are progressing satisfactorily.

We should perhaps mention that after the first two or three months it is a good idea to have the slower-growing species in a separate watering tray from those which are obviously faster-growing. Many of the succulents grow more quickly, and if removed from the others will be able to have more water which they can soak up faster.

Most of the young seedlings will by now have developed enough root to be fairly firm and not so easily dislodged in the soil. Regular watering from beneath by means of the damp tray helps them to send down deep roots and makes for healthy young plants. Overhead mist spraying, which can be given at quite frequent intervals in hot weather, keeps the atmosphere humid which is so helpful to plants at this stage. Always make sure that clean water is being used, to avoid the risk of accidentally introducing something harmful to the seedlings.

As they become sturdy little plants of some four, five or six months old, the bottom watering tray can be dispensed with. Stand the pots on the staging, spaced a little apart if room permits, and separate the quicker-growing seedlings from the slower species. You will find that this simplifies the overhead watering—*but the young plants must still be given some shade.*

The condition of the young plants can best be judged by their colour or general growth. So long as they are not trying to elongate unduly, and assuming the colour is not pallid as already mentioned, they are having the right amount of shade.

Treatment is now much easier. Watering need not be done regularly, and they can often be left without attention for several days. Weather conditions play a part in this, of course; in very hot weather which dries up all the moisture, some water is necessary.

When the young seedlings are growing well, they should be nicely firm and obviously making root as well as developing spines, etc. Old empty seed cases can be removed if they lie around the seedlings too thickly, since they can go mouldy and cause rot, particularly in a crowded seed pan. Provided the seedlings are firmly established, the empty cases can be lightly 'blown' off or picked off with tweezers. Seed cases still attached to the seedling, often found on one of the 'seed-leaves', do not need to be removed.

WINTERING THE YOUNG SEEDLINGS

There is no cause for alarm if cactus seedlings are rather crowded together in their original seed-pan. They will winter quite well and if kept just warm can continue growing, or what is important, making new root.

Continue watering on warm days until quite late in the year, and if you have warmth in your greenhouse small amounts of water can be given, perhaps at intervals of two to three weeks, throughout the first winter. Avoid this in very severe weather, however. If they dry out now, they will not be likely to harm.

If there is no proper warmth in the greenhouse, take the seedlings indoors, keep them in a light position during the daytime *but* well away from windows or unwarmed rooms at night. These conditions are actually very favourable to them at this stage.

If you have the bad luck to notice that a seedling has rotted at any stage of its growth (most likely due to too much water), pick it out with tweezers. This should not occur, but anyone can have an occasional mishap of this kind.

The seedlings are now nearing the end of their first season after germination, and in their second season can be treated as young plants in accordance with the general advice on cultivation given in the other chapters.

However, a little advice on the handling of small seedlings would not come amiss, as we know that it does sometimes cause problems. It is best to prick out into seed boxes or better still metal or plastic trays, which are 2–3 in. (5–7·5 cms) deep. These should have drainage holes. The soil mixture for plants has already been covered in an earlier chapter, so no further advice is necessary on that. When seedlings are to be pricked out it is preferable to water them the day before, so that seed soil is neither too dry nor too wet. If too dry the roots of the seedlings could be damaged, and if too wet it can make the job of pricking out more difficult.

A small pair of tweezers, such as those used by a philatelist, with spade ends are needed. With your tweezers, make a hole for each seedling, and then tap out each little pot of seedlings, so that each one can be easily lifted away from the others using the tweezers gently, so as not to crush the seedlings. The seedlings should then be put into the ready prepared hole down to the seed leaf position, and then gently firm it in with the tweezers. Freshly planted seedlings can then be lightly sprayed overhead, to settle them in completely.

DIAGRAMS ILLUSTRATING THE PROTECTION OF YOUNG SEEDLINGS

Fig. 11

Figs. 11 and 12.

These two sketches illustrate a simple way of protecting the young seedlings from direct sunshine, yet allowing a free flow of air when closed.

It is easy to get at the seedlings by lifting the muslin-covered lid. Any other method which is based upon these sketches can easily be assembled by the average beginner with little (if any) cost.

These sketches show the frame standing on a greenhouse staging through which some warmth from a small oil-heater can reach the bottom of the tray or dish on cold nights, should this be thought necessary.

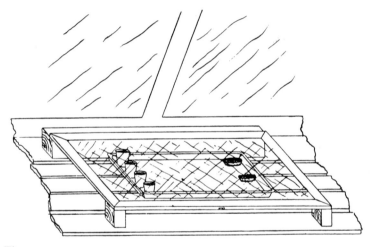

Fig. 12

SEEDLINGS

Figs. 13–21 have been included as a guide to the appearance of some varied seedlings, from 3–4 weeks of age to 6 months. One will find with many species that in the next 6 months little change occurs in their general appearance and size, such that further line drawings would have been superfluous.

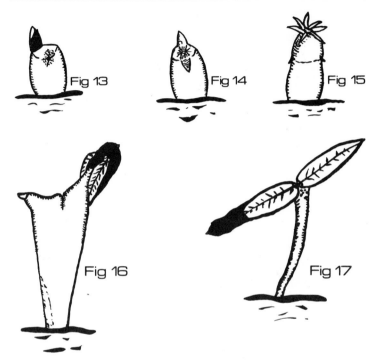

Fig. 13. A side view of a 3-week-old seedling of *Homalocephala texensis* shown at a magnification of × 4. Note the remains of the seed still attached to one of the seed leaves.

Fig. 14. A side view of a 2-month-old seedling of *Ferocactus wislizenii*, shown at a magnification of × 4. Note that the remains of the seed have long since disappeared and that the first rudimentary spines are just visible.

Fig. 15. A side view of a 3-month-old seedling of *Coryphantha radians*, shown at a magnification of × 4. Here the first spines are quite well formed, which will apply to most cacti seedlings of a similar age.

Fig. 16. A 3–4-week-old seedling of *Caralluma speciosa*, also shown side on, and again note that the remains of the seed are again visible on a seedling of this age. Magnification × 3.

Fig. 17. A side view of a 3-month-old seedling of *Ceropegia krainzii*, and with this the old seed remains are still attached to one of the seed leaves. However, it could easily be removed with a pair of tweezers. Magnification × 2.

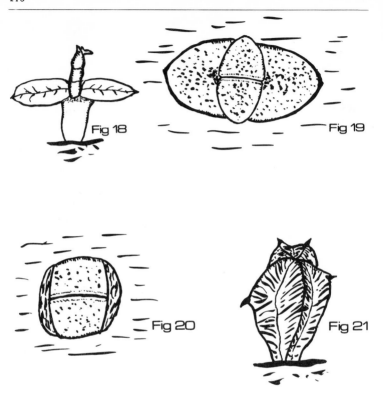

Fig. 18. Another side view shot of a 4-month-old seedling of *Ceropegia marnieri*, a trailing variety. At this stage the more mature trailing stem has started to appear from between the 'seed leaves'. Magnification × 2.

Fig. 19. A top view of a 3–4-month-old seedling of *Pleiospilos hilmari*, shown at a magnification of × 7. Already the second pair of leaves are well developed.

Fig. 20. A top view of a 6-month-old seedling of a *Lithops* species, shown at a magnification of × 6. In this case the original pair of 'leaves' have more or less dried up, leaving the seedling with a pair of leaves, as with a mature Lithops specimen.

Fig. 21. A side view of a 6-month-old seedling of *Astrophytum myriostigma*, shown at a magnification of × 2½. Here it is already a miniature of a mature specimen.

ACKNOWLEDGEMENTS

We would like specially to acknowledge the following Authorities already very well known in the field of 'Cacti and Succulents', not only for loaning us many fine illustrations to add to a few of our own for the plants in habitat section of this book, but also other help in connection with our work in 'The Exotic Collection'.

Mr Charles Glass, Editor of the 'Cactus and Succulent Journal' of the U.S.A.

Dr P. G. Corliss, of San Diego, California, well known also in other fields of horticulture as a lecturer.

Mr John Donald, Head of the Chemistry Department at Brighton College of Technology. Specialises in the Taxonomy of dwarf S. American cacti.

Mr Harry Hall, recently of the Botanical Gardens, Kirstenbosch, Cape Province, well known for his succulent exploration work throughout the southern half of Africa mainly over the last twenty years.

Mr John Lavranos, of Pretoria, Transvaal, another succulent explorer, particularly in recent years to remote spots such as the Audhali part of Aden, Socotra, and in South and East Africa.

Dr George Lindsay. Very well known authority on cacti, particularly those native to Baja California. Director of San Francisco Botanical Gardens.

Dr W. Rauh, of Heidelberg University in Germany, who has travelled very widely in search of succulents, including true cacti. His travels have taken him to most parts of Africa, Madagascar, South and Central America.

Mr Gordon Rowley. Lecturer in Botany at Reading University, particularly interested in the 'other succulents' with special emphasis on the Genus. Senecio.

Mr Gilbert Voss, of Encinitas, California, who has travelled widely in the southern States of the U.S.A., Mexico, including the remoter parts of Baja California.

Mr Charles Glass contributed Plate 325.

Dr P. G. Corliss contributed Plates 303 and 311.

Mr Harry Hall contributed Plates 315, 316, 319 and 323.

Mr John Lavranos contributed Plates 317 and 320.

Dr W. Rauh contributed Plates 301, 304, 305, 307, 308, 309, 310, 312, 313, 321 and 326.

Mr Gilbert Voss contributed Plates 306 and 324.

Dr George Lindsay contributed Plate 302.

All other illustrations in this book, including habitat Plates 314, 318 and 322, have been taken by the authors.

THE EXOTIC COLLECTION

Under the personal direction of
EDGAR LAMB and BRIAN M. LAMB

16 Franklin Road,
Worthing, Sussex,
BN13 2PQ, England.

EVERY MONTH The Exotic Collection sends its subscribers TWO NEW (previously unpublished) PHOTOGRAPHIC REFERENCE PLATES in COLOUR. (Size $8\frac{1}{2} \times 6$ in., 23 cm $\times$ 15 cm), with non-technical cultural notes, etc.

AN EIGHT PAGE 'Monthly Notes' (also illustrated in COLOUR). A total for one year of 24 'Plates' and 96 pages of Monthly Notes—minimum 72 pages of FULL COLOUR, plus articles and other cultural information. Some of these colour illustrations will cover two pages filmed by us here in the collection or in habitat.

OVERSEAS subscribers also receive an additional four-page Overseas Newsletter, containing specialised cultural information for varying countries, and this is usually issued bi-monthly.

ALL SUBSCRIBERS receive a seeds list of named species, some of which are free. Plants are also available to subscribers.

SUBSCRIPTION operates from January to December each year.

SUBSCRIPTION for ONE YEAR: £4.50 for the UK and Ireland.

SUBSCRIPTION for ONE YEAR: $10.00 (USA) or equivalent amount for all overseas countries.

INDEX